Polarization and Directional Effects in the Radiation from Plasmas

Online at: https://doi.org/10.1088/978-0-7503-6285-6

IOP Series in Plasma Physics

Series Editor

Richard Dendy

Culham Centre for Fusion Energy and the University of Warwick, UK

About the series

The IOP Plasma Physics ebook series aims at comprehensive coverage of the physics and applications of natural and laboratory plasmas, across all temperature regimes. Books in the series range from graduate and upper-level undergraduate textbooks, research monographs and reviews.

The conceptual areas of plasma physics addressed in the series include:

- Equilibrium, stability, and control.
- Waves: fundamental properties, emission, and absorption.
- Nonlinear phenomena and turbulence.
- Transport theory and phenomenology.
- Laser–plasma interactions.
- Non-thermal and suprathermal particle populations.
- Beams and non-neutral plasmas.
- High energy density physics.
- Plasma–solid interactions, dusty, complex, and non-ideal plasmas.
- Diagnostic measurements and techniques for data analysis.

The fields of application include:

- Nuclear fusion through magnetic and inertial confinement.
- Solar–terrestrial and astrophysical plasma environments and phenomena.
- Advanced radiation sources.
- Materials processing and functionalisation.
- Propulsion, combustion, and bulk materials management.
- Interaction of plasma with living matter and liquids.
- Biological, medical, and environmental systems.
- Low-temperature plasmas, glow discharges, and vacuum arcs.
- Plasma chemistry and reaction mechanisms.
- Plasma production by novel means.

A full list of titles published in this series can be found here: https://iopscience.iop.org/bookListInfo/iop-plasma-physics-series.

Polarization and Directional Effects in the Radiation from Plasmas

Eugene Oks
Physics Department, Auburn University, Auburn, USA

IOP Publishing, Bristol, UK

ISBN 978-0-7503-6285-6 (ebook)
ISBN 978-0-7503-6283-2 (print)
ISBN 978-0-7503-6286-3 (myPrint)
ISBN 978-0-7503-6284-9 (mobi)

DOI 10.1088/978-0-7503-6285-6

Version: 20240401

IOP ebooks

British Library Cataloguing-in-Publication Data: A catalogue record for this book is available from the British Library.

Published by IOP Publishing, wholly owned by The Institute of Physics, London

IOP Publishing, No.2 The Distillery, Glassfields, Avon Street, Bristol, BS2 0GR, UK

US Office: IOP Publishing, Inc., 190 North Independence Mall West, Suite 601, Philadelphia, PA 19106, USA

To Irisha with love

Contents

Author biography

Eugene Oks

Eugene Oks received his PhD degree from the Moscow Institute of Physics and Technology, and later the highest degree of Doctor of Sciences from the Institute of General Physics of the Academy of Sciences of the USSR by the decision of the Scientific Council led by the Nobel Prize winner, academician A M Prokhorov. According to the Statute of the Doctor of Sciences degree, this highest degree is awarded only to the most outstanding PhD scientists who founded a new research field of a great interest. Oks worked in Moscow (USSR) as the head of a research unit at the Center for Studying Surfaces and Vacuum, then—at the Ruhr University in Bochum (Germany) as an invited professor, and for the last 30 plus years—at the Physics Department of the Auburn University (USA) in the position of Professor. He conducted research in five areas: atomic and molecular physics, astrophysics, plasma physics, laser physics, and nonlinear dynamics. He founded/co-founded and developed new research fields, such as intra-Stark spectroscopy (new class of nonlinear optical phenomena in plasmas), masing without inversion (advanced schemes for generating/amplifying coherent microwave radiation), and quantum chaos (nonlinear dynamics in the microscopic world). He also developed a large number of advanced spectroscopic methods for diagnosing various laboratory and astrophysical plasmas—the methods that were then used and are used by many experimental groups around the world. He recently revealed that there are two flavors of hydrogen atoms, as proven by the analysis of atomic experiments; there is also a possible astrophysical proof—from observations of the anomalous absorption of the 21 cm radio line from the early Universe and from the observed, too smooth distribution of the dark matter in the Universe. He showed that dark matter or at least a part of it can be represented by the second flavor of hydrogen atoms. He published over 600 papers and 13 books, including the books *Plasma Spectroscopy: The Influence of Microwave and Laser Fields*, *Stark Broadening of Hydrogen and Hydrogenlike Spectral Lines in Plasmas: The Physical Insight*, *Breaking Paradigms in Atomic and Molecular Physics*, *Diagnostics of Laboratory and Astrophysical Plasmas Using Spectral Lineshapes of One-, Two, and Three-Electron Systems*, *Unexpected Similarities of the Universe with Atomic and Molecular Systems: What a Beautiful World*, 'Analytical Advances in Quantum and Celestial Mechanics: Separating Rapid and Slow Subsystems', *Advances in X-Ray Spectroscopy of Laser Plasmas*, *Simple Atomic and Molecular Systems: New Results and Applications*, *Advances in the Physics of Rydberg Atoms and Molecules*, *The Second Flavor of Hydrogen Atoms—The Leading Candidate for Dark Matter: Theoretical Discovery and the Proofs from Experiments and Astrophysical Observations*, *Recent Advances in Dark Matter, Dark Energy, Exoplanets, Flare Stars, White Dwarfs, and More*, and *Nonlinear Phenomena in the Radiation from Plasmas*. He is the Editor-in-Chief of the journals *Physics*

International and *International Review of Atomic and Molecular Physics*. He is a member of the Editorial Boards of six other journals: *Symmetry*, *American Journal of Astronomy and Astrophysics*, *Dynamics*, *Open Physics*, *Open Journal of Microphysics*, and *Current Physics*. He is a member of the Reviewers Board of the journal *Atoms*. He is also a member of the International Program Committees of the biannual series of conferences 'Spectral Line Shapes'.

IOP Publishing

Polarization and Directional Effects in the Radiation from Plasmas

Eugene Oks

Chapter 1

Introduction

In many practical situations, the radiation from plasmas occurs in the conditions where the spherical symmetry is broken to a lesser symmetry, such as, for instance, the axial symmetry. This happens in a variety of situations. One example is when there is a Quasimonochromatic Electric Field (QEF) in the plasma. The QEF can be intrinsic, such as, e.g., the Langmuir waves, or extrinsic, such as e.g., the laser field. The laser field can be linearly-polarized (just as the Langmuir waves) or circularly-polarized—in both cases the spherical symmetry is reduced to the axial symmetry. Alternatively, the laser field can be elliptically-polarized, in which case there is not even the axial symmetry.

In any of these cases, there exists a preferred direction in the plasma: either the direction of the linearly-polarized field, or the direction perpendicular to the field polarization plane (in the cases of circular or elliptical polarization). Due to the existence of the preferred direction, there occur polarization and directional effects in the radiation from plasmas.

Some of these effects had an initial coverage in book [1] of 1995 titled *Plasma Spectroscopy: The Influence of Microwave and Laser Fields*. Also, there is book [2] of 2007 titled *Plasma Polarization Spectroscopy*. However, in the intervening 17 years, lots of very significant advances have been made in those topics of the latter book. Besides, book [2] of 2007 did not cover polarization and directional effects in the radiation from plasmas containing the QEF. Meanwhile, in the subsequent 17 years there occurred lots of theoretical and experimental studies of the radiation from plasmas containing the QEF and those studies revealed important new phenomena.

The present book not only has theoretical importance, but also practical significance. It has applications (described in the book) to numerous areas of diagnostics of laboratory and astrophysical plasmas, including (but not limited to) laser–plasma interactions and plasma devices for magnetically-controlled thermonuclear fusion. Together with book [3] of 2023 *Nonlinear Phenomena in the*

doi:10.1088/978-0-7503-6285-6ch1

Radiation from Plasmas, the present book constitutes the second part of the research dilogy providing the comprehensive description of various aspects of the radiation from plasmas for the research community dealing with atomic processes in plasmas.

The distinctive feature of the present book are as follows:

- All the nonlinear phenomena within the scope of the book are revealed by analytical methods, thus providing the physical insight that simulations lack;
- One of the analytical methods employs quasienergy states, which is an advanced formalism facilitation of solutions and better understanding of many problems in this research area;
- Various polarization and directional effects in the radiation from plasmas, presented in numerous research papers, each covering only one aspect, are collected 'under one roof', that is, under the cover of this book;
- This book can enable readers to comprehend the interconnections between different polarization and directional effects in the radiation from plasmas and to engage in the corresponding research with the necessary physical background.

Chapter 2 presents the polarization and directional effects in the radiation of satellites of *hydrogenic* spectral lines from plasmas and their applications. It describes various types of satellites of hydrogenic spectral lines emitted by plasmas. The types of the satellites depend on the type of the QEF (extrinsic or intrinsic to plasmas) that causes the satellites: for example, on the number of modes of the QEF and on the polarization of the QEF. As a result, the satellites will be polarized; plus, their shape will depend on the direction of the observation. The coverage is followed by the discussion of the practical applications of these effects for spectroscopic diagnostics of various plasmas based on the employment of hydrogenic spectral lines. The relevant literature references are [4–13].

Chapter 3 describes various types of satellites of *non-hydrogenic* spectral lines emitted by plasmas. From the theoretical viewpoint, the focus is on the analytical methods beyond the perturbation theory. The utilization of these methods reveals the dependence of the satellites on the polarization and on the direction of the observation. The relevant literature references are [14–23].

Chapter 4 depicts the emergent phenomenon of the Langmuir-wave-caused 'dips' in spectral line profiles, where the word 'dip', used for brevity, refers to a highly-localized structures in the line profile—the structures consisting of a local minimum of the intensity surrounded by two 'bumps' (peaks). This emergent phenomenon originates from *multifrequency nonlinear dynamic resonances* that involve the coupling of the following three plasma subsystems: the QEF, the quasistatic electric field, and the radiating atom or ion. The focus is on the polarization of the 'dips' and on their dependence on the direction of observations. The coverage is followed by the discussion of the practical applications of these features of the Langmuir dips for spectroscopic diagnostics of various laboratory and astrophysical plasmas in view of the fact that the Langmuir-wave-caused 'dips' were confirmed in a large number of experiments by various experimental groups working at different plasma machines, as well as in astrophysical observations. The relevant literature references are [24–44].

Chapter 5 is devoted to the polarization and directional effects in the *non-resonant* coupling of the monochromatic and quasistatic electric fields in plasmas. In particular, it describes the partial or complete suppression of the quasistatic part of the plasma microfield by the OEF: namely, the suppression of the components of the quasistatic electric field that are perpendicular to the linearly-polarized OEF. The coverage is followed by the discussion of the practical applications of these effects for spectroscopic diagnostics of plasmas in magnetic fusion machines. The relevant literature references are [45, 46].

Chapter 6 describes the polarization and directional effects in the *anisotropic* Stark broadening of hydrogen spectral lines due to the motion of the radiating ions. For a radiating atom moving with a constant velocity, in the reference frame of the radiating atom, the plasma appears anisotropic. Consequently, the spectral line profiles depend both on the velocity of the radiating atom and on the direction of observation. The corresponding changes in the spectral line profiles are especially significant in the case where the dynamical Stark broadening is caused by plasma ions that are much heavier than the radiating atom. The anisotropic character of the Stark broadening leads to the *narrowing* of the line profile. The relevant literature references are [47–54].

Chapter 7 depicts the polarization and directional effects in the radiation of the plasma-based *x-ray lasers*. One avenue in designing soft x-ray lasers is grounded on the recombination pumping of hydrogen-like ions—the ions from which all electrons, except one, were removed by the optical field ionization. The gain of x-ray lasers is controlled by the product of the oscillator strength for the lasing transition and the halfwidth of the lasing line. In the relevant literature there was demonstrated the possibility to significantly diminish the Stark width of the hydrogen-like spectral lines by the application of a high frequency electric field: the latter modifies the coupling of the electric microfield of a plasma with the hydrogen-like ion. In particular, it was revealed that by applying a linearly-polarized field of an optical laser it is possible—for some x-ray lasing hydrogenic spectral lines—to substantially narrow the profile of the absorption coefficient and in this way to increase the gain of the x-ray laser. In the literature it was also considered how the elliptically-polarized intense field of an optical laser (EPIFOL) affects the Stark broadening of some x-ray spectral lines of hydrogen-like ions. It was demonstrated that the employment of the EPIFOL has the following two advantages. First, the gain of some x-ray lasing transitions will be substantially increased. Second, it would open up the possibility to create a tunable x-ray laser—the laser tunable in a broad range of frequencies. The relevant literature references are [8, 55–66].

Chapter 8 is devoted to the Stark broadening of spectral lines by a specific kind of moving emitters. First, it describes the Stark broadening of spectral lines emitted by nonrelativistic neutral beams. Then it describes the Stark broadening of spectral lines by relativistic electron beams. In all of the above cases, the outcome depends on the polarization and on the direction of the observation. The coverage is followed by the discussion of the practical applications of these effects for spectroscopic diagnostics of magnetic fusion plasmas in view of the fact that the neutral beams and the relativistic electron beams are used for heating the magnetic fusion plasmas, and that the relativistic electron beams can also occur intrinsically in these plasmas due to the phenomenon of run-away electrons. The relevant literature references are [67–84].

Chapter 9 describes the polarization and directional effects in the Stark broadening of spectral lines in strongly-magnetized plasmas. In particular, it covers the polarization and directional differences between the Stark broadening of π- and σ-components of hydrogenic spectral lines in strongly-magnetized plasmas. For example, the π-component of the Ly-alpha line turns out to be unaffected by the magnetic field, but the σ-components turn out to be strongly affected by the magnetic field. The π-component of the Lyman-alpha line has also another remarkable feature. In the case of the dominance of the Zeeman effect over the Stark effects, the broadening of the π-component of the Lyman-alpha line is essentially controlled by the Stark effect: practically no dependence on the magnetic field. This is a clear distinction from the σ-component of the Lyman-alpha line or from any component of any other spectral line of hydrogenic atoms or ions. The chapter also depicts the Lorentz–Doppler broadening of hydrogen or deuterium spectral lines. The entire coverage is followed by the discussion of the practical applications of these effects for spectroscopic diagnostics of various strongly-magnetized plasmas, especially the magnetic fusion plasmas and the plasmas in the atmospheres of white dwarfs. The relevant literature references are [85–98].

Chapter 10 is devoted to the polarization and directional effects in the laser-induced fluorescence from plasmas and their applications. It covers various polarization and directional effects in the area of active (rather than passive) methods of plasma spectroscopy. The effects under consideration occur while using the laser-induced fluorescence from plasmas as the diagnostic method. This method has the advantage over the passive spectroscopic methods: it allows obtaining the diagnostic information with the spatial and temporal resolutions. The coverage is followed by the discussion of the practical applications of these effects for laser-aided spectroscopic diagnostics of the QEF in various plasmas. The relevant literature references are [99–103].

References

[1] Oks E 1995 *Plasma Spectroscopy: The Influence of Microwave and Laser Fields* (Berlin: Springer)

[2] Fujimoto T and Iwamae A 2007 *Plasma Polarization Spectroscopy* (Berlin: Springer)

[3] Oks E 2023 *Nonlinear Phenomena in the Radiation from Plasmas* (Bristol: IOP Publishing)

[4] Blochinzew D I 1933 *Phys. Z. Sow. Union* **4** 501

[5] Renner O, Peyrusse O, Sondhauss P and Förster E 2000 *J. Phys. B: At. Mol. Opt. Phys.* **33** L151

[6] Lifshitz E V and Phys S 1968 *JETP* **26** 570

[7] Weinheimer J, Oks E, Clothiaux E J, Schulz A and Svidzinski V 1998 *IEEE Trans. Plasma Sci.* **26** 1239

[8] Oks E and Gavrilenko V P 1983 *Opt. Commun.* **46** 205

[9] Volod'ko D A, Gavrilenko V P and Oks, E.A. E 1987 *18th Int. Conf. on Phenomena in Ionized Gases (Swansea, UK)* p 604

[10] Ishimura T 1967 *J. Phys. Soc. Jpn.* **23** 422
[11] Lisitsa V S 1971 *Opt. Spectrosc.* **31** 468
[12] Zel'dovich J B 1967 *Sov. Phys. JETP* **24** 1006
[13] Ritus V I 1967 *Sov. Phys. JETP* **24** 1041
[14] Baranger M and Mozer B 1961 *Phys. Rev.* **123** 25
[15] Cooper W S and Ringler H 1969 *Phys. Rev.* **179** 226
[16] Oks E and Gavrilenko V P 1983 *Sov. Tech. Phys. Lett.* **9** 111
[17] Oks E 2017 *Diagnostics of Laboratory and Astrophysical Plasmas Using Spectral Lines of One-, Two-, and Three-Electron Systems* (Hackensack, NJ: World Scientific)
[18] Brizhinev M P, Gavrilenko V P, Egorov S V, Eremin B G, Kostrov A V, Oks E and Shagiev Y M 1983 *Sov. Phys. JETP* **58** 517
[19] Perelman N F and Mosyak A A 1989 *Sov. Phys. JETP* **69** 700
[20] Gavrilenko V P and Oks E 1982 *Proc. Int. Conf. on Plasma Physics (Geteborg, Sweden)* p 353
[21] Gavrilenko V P and Oks E 1983 *Sov. J. Quantum Electron.* **13** 1269
[22] Gavrilenko V P and Oks E 1989 *Opt. Commun.* **69** 384
[23] Gavrilenko V P and Oks E 1986 *Proc. 13th Summer School and Int. Symp. on Physics of Ionized Gases (Sibenik, Yugoslavia)* p 393
[24] Gavrilenko V P and Oks E 1987 *Sov. Phys. J. Plasma Phys.* **13** 22
[25] Oks E, Böddeker S and Kunze, HJ H-J 1991 *Phys. Rev.* A **44** 8338
[26] Gavrilenko V P and Oks E 1981 *Sov. Phys. JETP* **53** 1122
[27] Oks E, Böddeker S and Kunze H-J 1991 *Phys. Rev.* A **44** 8338
[28] Dalimier E, Oks E and Renner O 2014 *Atoms* **2** 178
[29] Dalimier E, Ya A, Faenov , Oks E, Angelo P, Pikuz T A, Fukuda Y *et al* 2017 *J. Phys.: Conf. Ser.* **810** 012004
[30] Dalimier E, Oks E and Renner O 2017 *AIP Conf. Proc.* **1811** 190003
[31] Dalimier E and Oks E 2018 *Atoms* **6** 60
[32] Oks E, Dalimier E and Angelo P 2019 *Spectrochim. Acta* B **157** 1.
[33] Dalimier E, Oks E and Angelo P 2020 *Int. Rev. At. Mol. Phys.* **11** 1
[34] Belyaev V S, Krainov V P, Lisitsa V S and Matafonov A P 2008 *Phys.-Usp.* **51** 793
[35] Belyaev V S and Matafonov A P 2011 *Femtosecond-Scale Optics* ed A Andreev (Shanghai: InTech) vol 70 p 87
[36] Tatarakis M, Gopal A, Watts I, Beg F N, Dangor A E, Krushelnik K *et al* 2022 *Phys. Plasmas* **9** 2244
[37] Tatarakis M, Watts I, Beg F N, Clark E L, Dangor A E, Gopal A *et al* 2002 *Nature* **415** 280
[38] Kato S, Nakamura T, Mima K, Sentoku Y, Nagatomo H and Owadano Y 2004 *J. Plasma Fusion Res.* **6** 658
[39] Perogaro F, Bulanov S V, Califano F, Zh. Esirkepov T, Lontano M, Meyer-ter-Vehn J *et al* 1997 *Plasma Phys. Control. Fusion* **38** B26
[40] Singh M, Gopal K and Gupta D 2016 *Phys. Lett.* A **380** 1437
[41] Liseykina T V, Popruzhenko S V and Macchi A 2016 *New J. Phys.* **18** e072001
[42] Santos J J, Bailly-Crandvaux M, Ehret M, Arefiev A V, Batani D, Beg F N *et al* 2018 *Phys. Plasmas* **25** e056705
[43] Oks E and Sholin G V 1977 *Opt. Spectrosc.* **42** 434
[44] Zhuzhunashvili A I and Oks E 1977 *Sov. Phys. JETP* **46** 1122

[45] Gavrilenko V P and Oks E 1985 *Proc. 17th Int. Conf. on Phenomena in Ionized Gases (Budapest)* p 1081
[46] Gavrilenko V P, Oks E and Rantsev-Kartinov V A 1986 *JETP Lett.* **44** 404
[47] Seidel J 1979 *Z. Naturforsch.* **34a** 1389
[48] Griem H R 1974 *Spectral Line Broadening by Plasmas* (New York: Academic)
[49] Derevianko A and Oks E 1995 *J. Quant. Spectrosc. Radiat. Transfer* **54** 137
[50] Ispolatov Y and Oks E 1994 *J. Quant. Spectrosc. Radiat. Transfer* **51** 129
[51] Oks E 2006 *Stark Broadening of Hydrogen and Hydrogenlike Spectral Lines in Plasmas: The Physical Insight* (Oxford: Alpha Science International)
[52] Derevianko A and Oks E 1994 *Phys. Rev. Lett.* **73** 2059
[53] Stehle C and Feautrier N 1984 *J. Phys.* B **17** 1477
[54] Gaisinsky I M and Oks E A 1985 *J. Phys.* B **18** 1449
[55] Nagata Y, Midorikawa K, Kubodera S, Obara M, Tashiro H and Toyoda K 1993 *Phys. Rev. Lett.* **71** 3774
[56] Donnelly T D, Da Silva L, Lee R W, Mrowka S, Hofer M and Falcone R W 1996 *J. Opt. Soc. Am.* **B13** 185
[57] Korobkin D V, Nam C H, Suckewer S and Goltsov A 1996 *Phys. Rev. Lett.* **77** 5206
[58] Gavrilenko V P and Oks E 1989 *Proc. 19th Int. Conf. on Phenomena in Ionized Gases (Belgrade, Yugoslavia)* p 354
[59] Oks E 2000 *J. Phys. B: At. Mol. Opt. Phys.* **33** L801
[60] Gavrilenko V P and Oks E 2004 *Eur. Phys. J.* D **28** 253
[61] Krylov N N and Bogoliubov N N 1947 *Introduction to Non-linear Mechanics* (Princeton, NJ: Princeton University Press)
[62] Bogoliubov N N and Mitropolskii Y M 1961 *Asymptotic Methods in the Theory of Nonlinear Oscillations* (New York: Gordon and Breach)
[63] Brissaud A and Frisch U 1971 *J. Quant. Spectrosc. Radiat. Transfer* **11** 1767
[64] Golosnoy I O 1993 *Matematicheskoe Modelirovanie [Math. Modeling]* **5** 11 (in Russian)
[65] Iglesias C A and Lebowitz J L 1984 *Phys. Rev.* A **30** 2001
[66] Iglesias C A, DeWitt H E, Lebowitz J L, MacGowan D and Hubbard W B 1985 *Phys. Rev.* A **31** 1698
[67] Brissaud A, Goldbach C, Léorat J, Mazure A and Nollez G 1976 *J. Phys. B: At. Mol. Phys.* **9** 1129
[68] Guenot D *et al* 2017 *Nat. Photonics* **11** 293
[69] Kurkin S A, Hramov A E and Koronovskii A A 2013 *Appl. Phys. Lett.* **103** 043507
[70] de Jagher P C, Sluijter F W and Hopman H J 1988 *Phys. Rep.* **167** 177
[71] Decker J, Hirvijoki E, Embreus O, Peysson Y, Stahl A, Pusztai I and Fülöp T 2016 *Plasma Phys. Control. Fusion* **58** 025016
[72] Smith H, Helander P, Eriksson L-G, Anderson D, Lisak M and Andersson F 2006 *Phys. Plasmas* **13** 102502
[73] Minashin P V, Kukushkin A B and Poznyak V I 2012 *EPJ Web Conf.* **32** 01015
[74] Kurzan B, Steuer K-H and Suttrop W 1997 *Rev. Sci. Instrum.* **68** 423
[75] Ide S *et al* 1989 *Nucl. Fusion* **29** 1325
[76] Oks E and Sanders P 2018 *J. Phys. Commun.* **2** 015030
[77] Rosato J, Pandya S P, Logeais C, Meireni M, Hannachi I, Reichle R, Barnsley R, Marandet Y and Stamm R 2017 *AIP Conf. Proc.* **1811** 110001
[78] Landau L D and Lifshitz E M 1971 *The Classical Theory of Fields* (Oxford: Pergamon)

[79] Hemsworth A S *et al* 2017 *New J. Phys.* **19** 025005
[80] Demkov Y, Monozon B and Ostrovskii V 1970 *Sov. Phys. JETP* **30** 775
[81] Kukushkin A B 2023 Private communication.
[82] Watts C *et al* 2013 *Nucl. Instrum. Methods Phys. Res.* A **720** 7
[83] Yatsuka E *et al* 2013 *J. Instrum.* **8** C12001
[84] Kukushkin A B, Kukushkin A S, Lisitsa V S, Neverov V S, Pshenov A A and Shurygin V A 2021 *Plasma Phys. Control. Fusion* **63** 035025
[85] Derevianko A and Oks E 1997 *Rev. Sci. Instrum* **68** 998
[86] Wroblewski D 1992 *AIP Conf. Proc Atomic Processes in Plasmas* **No. 257** (New York: AIP) p 121
[87] Abramov V A and Lisitsa V S 1977 *Sov. J. Plasma Phys.* **3** 451
[88] Bychkov S S, Ivanov R S and Stotskii G I 1987 *Sov. J. Plasma Phys.* **13** 769
[89] Oks E 2023 *Int. Rev. Atom. Mol. Phys.* **14** 43
[90] Sholin G V and Oks E 1973 *Sov. Phys. Doklady* **18** 254
[91] Oks E and Sholin G V 1976 *Sov. Phys. Tech. Phys.* **21** 144
[92] Oks E 2015 *J. Quant. Spectr. Radiat. Transfer* **156** 24
[93] Welch B L, Griem H R, Terry J, Kurz C, LaBombard B, Lipschultz B, Marmar E and McCracken J 1995 *Phys. Plasmas* **2** 4246
[94] Brooks N H, Lisgo S, Oks E, Volodko D, Groth M, Leonard A W and Team DIII-D 2009 *Plasma Phys. Rep.* **35** 112
[95] Oks E 2012 *Atomic Processes in Basic and Applied Physics* ed V Shevelko and H Tawara (Heidelberg: Springer) ch 15
[96] Oks E, Bengtson R D and Touma J 2000 *Contrib. Plasma Phys.* **40** 158
[97] Feldman U and Doschek G A 1977 *Astrophys. J.* **212** 913
[98] Galushkin Yu I 1970 *Sov. Astron. Zh.* **14** 301
[99] Sanders P and Oks E 2017 *J. Phys. Commun.* **1** 055011
[100] Gavrilenko V P and Oks E 2011 *Int. Rev. Atom. Mol. Phys.* **2** 35
[101] Delone N B and Krainov V P 1985 *Atoms in Strong Light Fields* (Berlin: Springer)
[102] Gavrilenko V P and Oks E 1995 *Phys. Rev. Lett.* **74** 3796
[103] Gavrilenko V P and Oks E 1995 *J. Phys. B: At. Mol. Opt. Phys.* **28** 1433

Chapter 2

Polarization and directional effects in the radiation of satellites of hydrogenic spectral lines from plasmas and their applications

2.1 Satellites under the one-dimensional one-mode monochromatic electric field

In this section we analyze a hydrogen atom or a hydrogenlike ion of a nuclear charge Z subjected to a linearly-polarized field $\mathbf{E}_0 \cos \omega t$, such as for instance, a laser field. Based on the results by Blochinzew [1], the profile of a Stark component of the spectral line can be written as follows (in terms of the scaled dimensionless detuning $\Delta\omega/\omega$ from the frequency ω_0 of the spectral line):

$$S_{\text{profile}}(\Delta\omega/\omega) = \sum_{p=-\infty}^{\infty} \left[J_p(X\varepsilon)\right]^2 \delta(\Delta\omega/\omega - p) \tag{2.1}$$

In equation (2.1), $J_p(X\varepsilon)$ denotes the Bessel functions, where in their arguments

$$\varepsilon = 3\hbar E_0/(2Zm_e e\omega), \quad X = (nq)_{\text{upper}} - (nq)_{\text{lower}}, \quad q = (n_1 - n_2). \tag{2.2}$$

In equation (2.2), e and m_e are the electron charge and mass, respectively; the parabolic quantum numbers have the standard notations n_1 and n_2, while the principal quantum number is denoted by n. As for the subscripts 'lower' and 'upper', they correspond to the lower and upper energy levels, between which the radiative transition occurs.

The normalization of the profile $S(\Delta\omega/\omega)$ is:

$$\int_{-\infty}^{\infty} S_{\text{profile}}(\Delta\omega/\omega) d(\Delta\omega/\omega) = 1. \tag{2.3}$$

doi:10.1088/978-0-7503-6285-6ch2

The profile consists of satellites at the locations $p\omega$ counted from the unperturbed frequency ω_0. Here p is any integer (positive or negative), while $p = 0$ labels the main line.

The profile of a multicomponent hydrogenic spectral line can be represented as follows—according to section 3.1 of book [2]:

$$S(\Delta\omega/\omega) = \sum_{p=-\infty}^{+\infty} I(p, \varepsilon)\, \delta(\Delta\omega/\omega) - p), \tag{2.4}$$

$$I(p, \varepsilon) = \left[f_0(\theta)\delta_{p0} + 2\sum_{k=1}^{k_{\max}} f_k(\theta) J_p^2(X_k\varepsilon) \right] / (f_0 + 2\Sigma f_k).$$

In equation (2.4), θ is the angle of the observation with respect to the vector-amplitude $\mathbf{E}_0$ of the electric field, f_k is the intensity of the lateral Stark component having the number $k = 1, 2, \ldots, k_{\max}$, while f_0 is the total intensity of all central Stark components.

Our focus is on the dependence of the profile on the angle of the observation θ. Below we illustrate the angular dependence of calculated profiles of the hydrogenic spectral lines that are the most useful for spectroscopic diagnostics of plasmas. For obtaining continuous profiles we assigned to each satellite the Lorentzian shape of the half width at half maximum equal to $\omega/4$. The scaled intensity S is calculated according to equation (2.4) versus the scaled dimensionless detuning

$$\delta = \Delta\omega/\omega. \tag{2.5}$$

Let us start from the Lyman-beta line. Figure 2.1 shows its profile for the scaled laser amplitude $\varepsilon = 0.5$ for four different observation angles: $\pi/2$ (solid line), $\pi/3$ (dash-dotted line), $\pi/6$ (dotted line), and 0 (dashed line).

Figure 2.2 demonstrates the calculated profiles of the Lyman-beta line for the scaled laser amplitude $\varepsilon = 1$ for four different observation angles: $\pi/2$ (solid line), $\pi/3$ (dash-dotted line), $\pi/6$ (dotted line), and 0 (dashed line).

Figure 2.3 displays the calculated profiles of the Lyman-beta line for the scaled laser amplitude $\varepsilon = 2$ for four different observation angles: $\pi/2$ (solid line), $\pi/3$ (dash-dotted line), $\pi/6$ (dotted line), and 0 (dashed line).

Figure 2.4 shows the calculated profiles of the Lyman-beta line for the scaled laser amplitude $\varepsilon = 4$ for four different observation angles: $\pi/2$ (solid line), $\pi/3$ (dash-dotted line), $\pi/6$ (dotted line), and 0 (dashed line).

Figure 2.5 demonstrates the calculated profiles of the Lyman-beta line for the scaled laser amplitude $\varepsilon = 8$ for four different observation angles: $\pi/2$ (solid line), $\pi/3$ (dash-dotted line), $\pi/6$ (dotted line), and 0 (dashed line).

Figure 2.6 displays the calculated profiles of the Lyman-beta line for the scaled laser amplitude $\varepsilon = 16$ for four different observation angles: $\pi/2$ (solid line), $\pi/3$ (dash-dotted line), $\pi/6$ (dotted line), and 0 (dashed line).

From figures 2.1–2.6 one can see the following. First, as the observation angle θ decreases from $\pi/2$ to 0, the profile narrows: its full width at half maximum

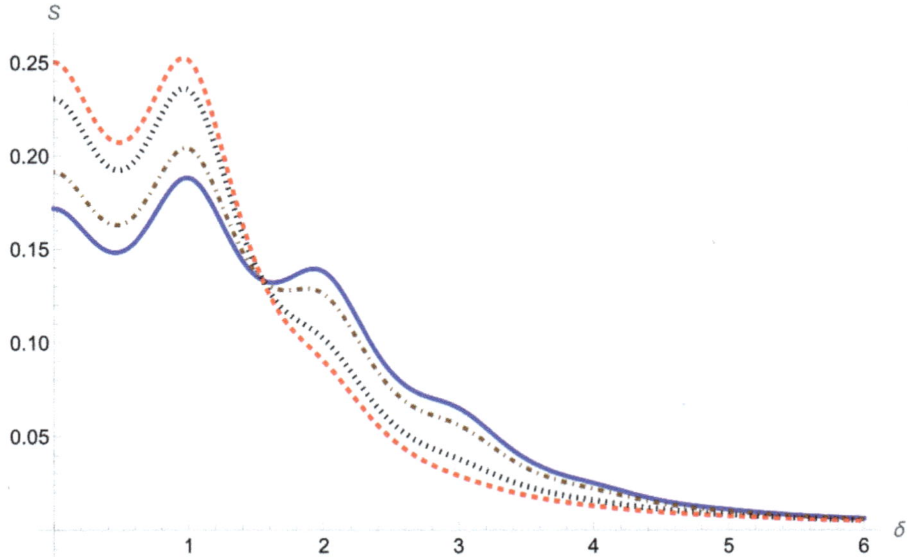

Figure 2.1. Calculated profiles of the Lyman-beta line for the scaled laser amplitude $\varepsilon = 0.5$ for four different observation angles: $\pi/2$ (solid line), $\pi/3$ (dash-dotted line), $\pi/6$ (dotted line), and 0 (dashed line).

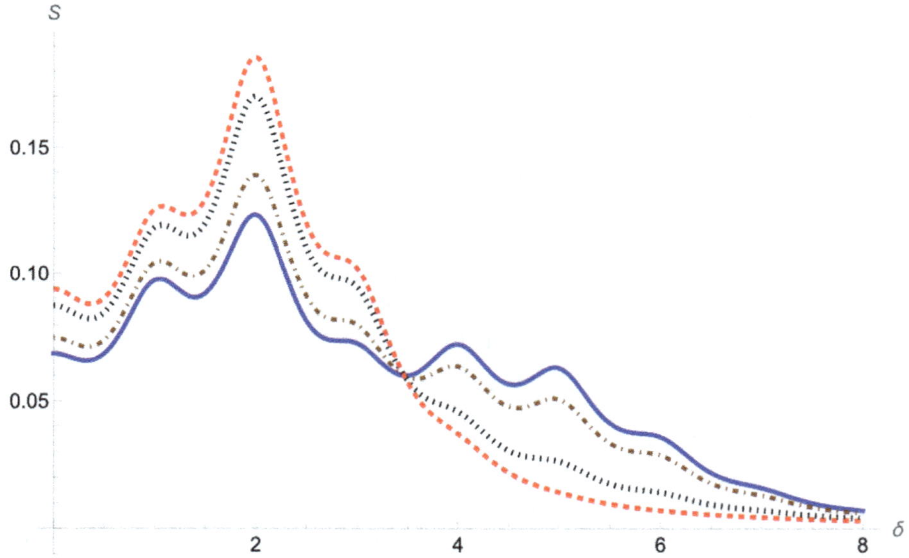

Figure 2.2. Calculated profiles of the Lyman-beta line as in figure 2.1, but for the scaled laser amplitude $\varepsilon = 1$.

decreases. This effect is more pronounced for relatively small values of the scaled laser amplitude ε.

Second, as the scaled laser amplitude ε increases, the profiles exhibit more and more structures. At the same time, the primary maximum of the profiles shifts to larger distances from the unperturbed position of the Lyman-beta line.

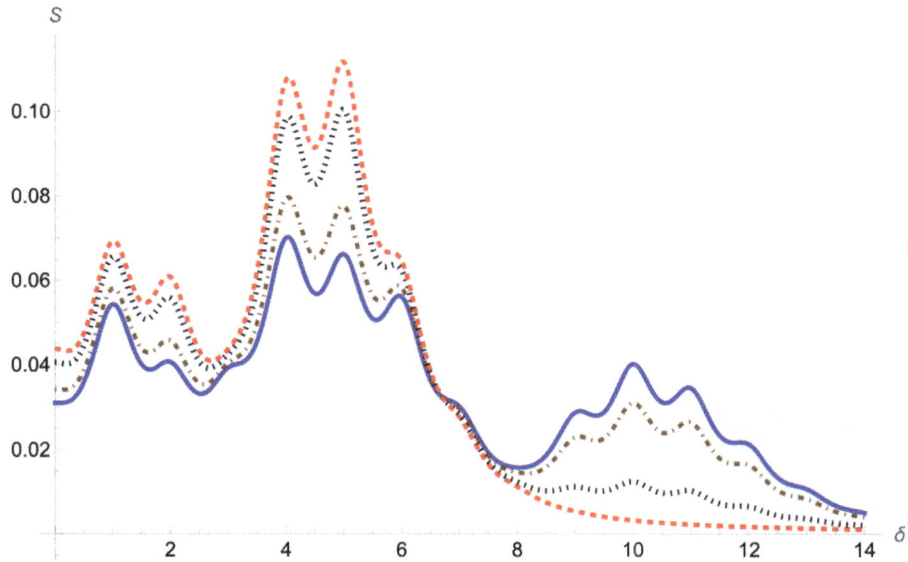

Figure 2.3. Calculated profiles of the Lyman-beta line as in figure 2.1, but for the scaled laser amplitude $\varepsilon = 2$.

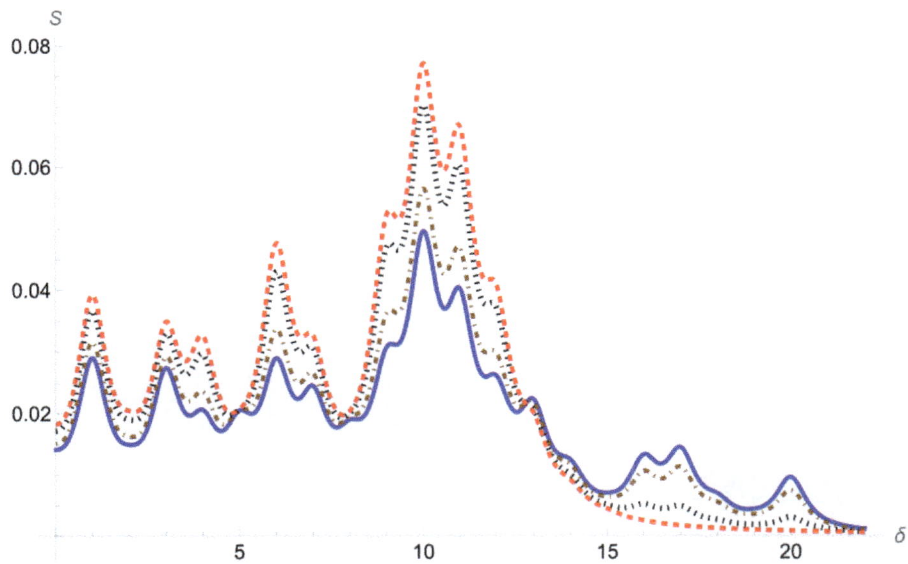

Figure 2.4. Calculated profiles of the Lyman-beta line as in figure 2.1, but for the scaled laser amplitude $\varepsilon = 4$.

Now we proceed to studying the angular dependence of the Lyman-delta profiles. Figure 2.7 shows its profiles for the scaled laser amplitude $\varepsilon = 0.125$ for four different observation angles: $\pi/2$ (solid line), $\pi/3$ (dash-dotted line), $\pi/6$ (dotted line), and 0 (dashed line).

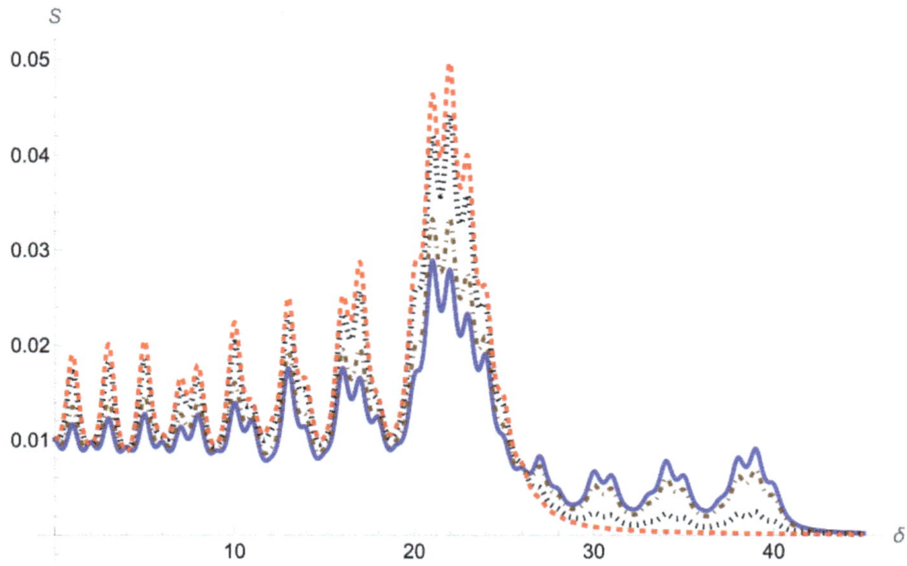

Figure 2.5. Calculated profiles of the Lyman-beta line as in figure 2.1, but for the scaled laser amplitude $\varepsilon = 8$.

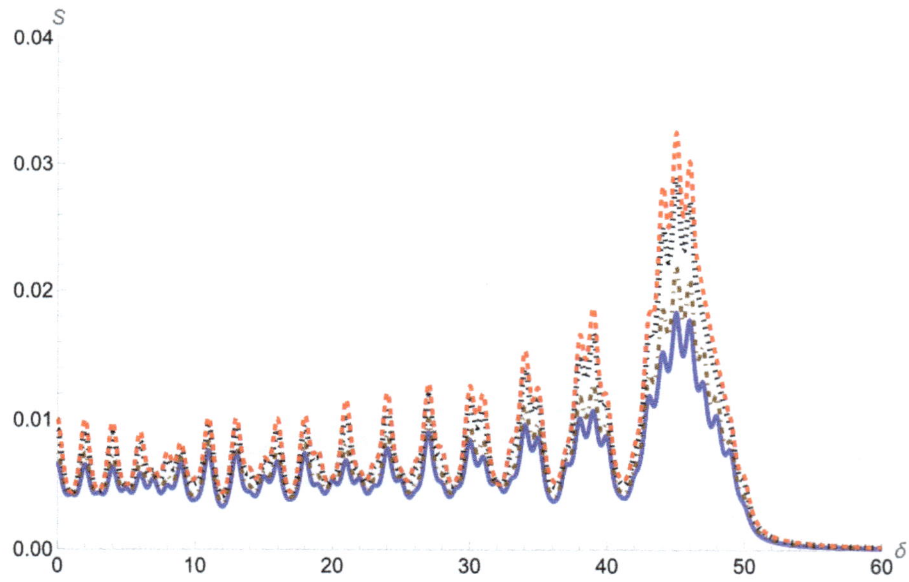

Figure 2.6. Calculated profiles of the Lyman-beta line as in figure 2.1, but for the scaled laser amplitude $\varepsilon = 16$.

Figure 2.8 demonstrates the calculated profiles of the Lyman-beta line for the scaled laser amplitude $\varepsilon = 0.25$ for four different observation angles: $\pi/2$ (solid line), $\pi/3$ (dash-dotted line), $\pi/6$ (dotted line), and 0 (dashed line).

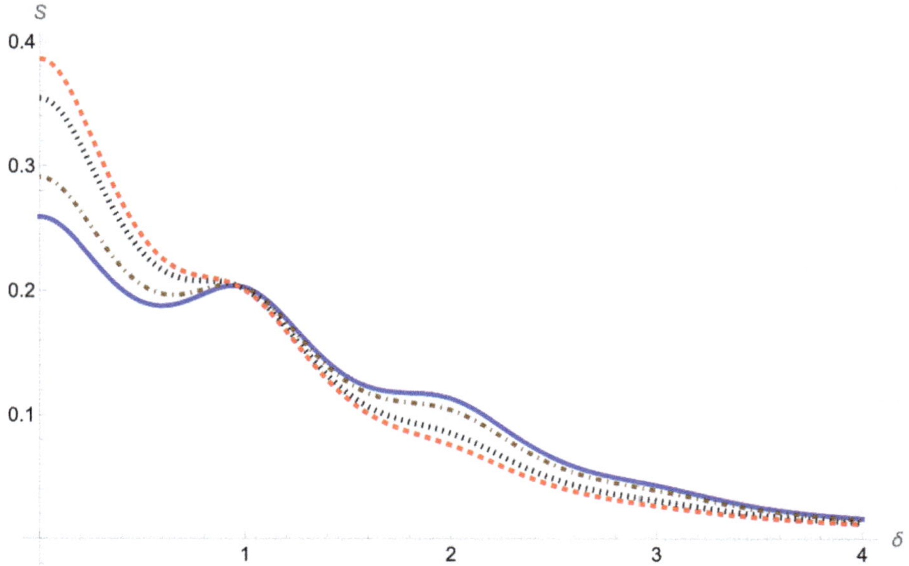

Figure 2.7. Calculated profiles of the Lyman-delta line for the scaled laser amplitude $\varepsilon = 0.125$ for four different observation angles: $\pi/2$ (solid line), $\pi/3$ (dash-dotted line), $\pi/6$ (dotted line), and 0 (dashed line).

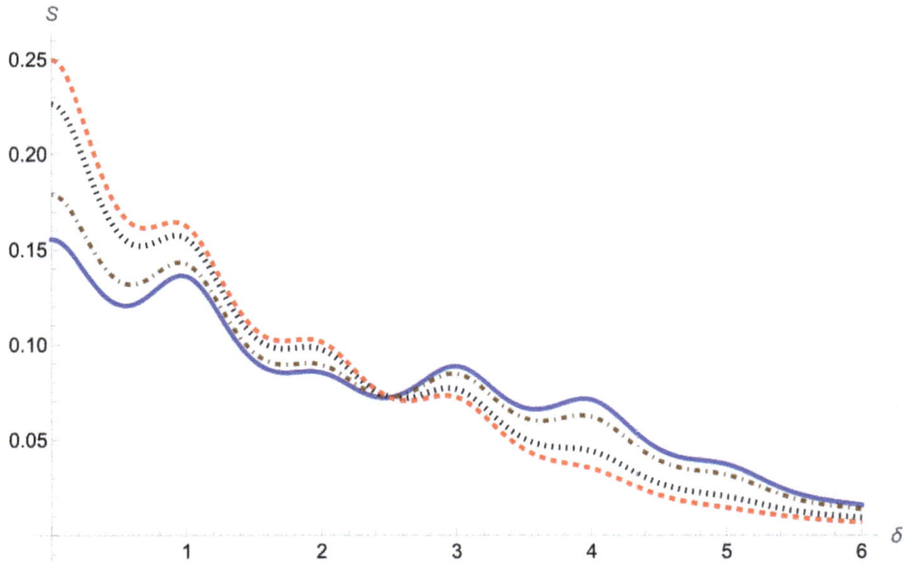

Figure 2.8. Calculated profiles of the Lyman-delta line as in figure 2.7, but for the scaled laser amplitude $\varepsilon = 0.25$.

Figure 2.9 displays the calculated profiles of the Lyman-beta line for the scaled laser amplitude $\varepsilon = 0.5$ for four different observation angles: $\pi/2$ (solid line), $\pi/3$ (dash-dotted line), $\pi/6$ (dotted line), and 0 (dashed line).

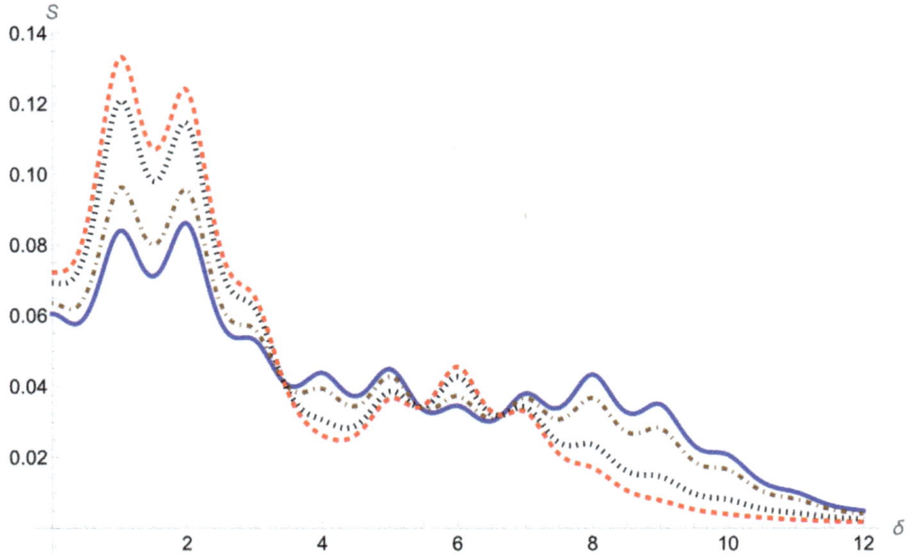

Figure 2.9. Calculated profiles of the Lyman-delta line as in figure 2.7, but for the scaled laser amplitude $\varepsilon = 0.5$.

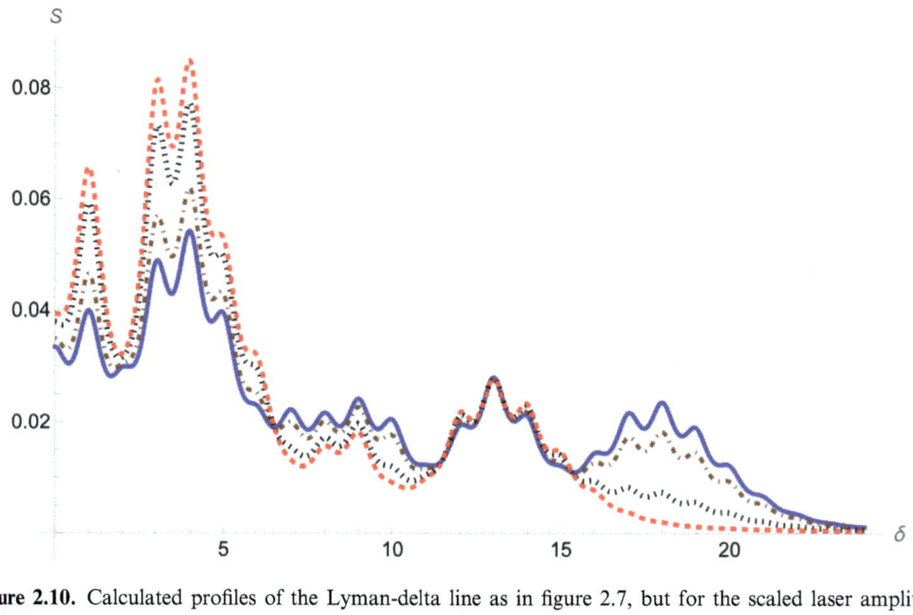

Figure 2.10. Calculated profiles of the Lyman-delta line as in figure 2.7, but for the scaled laser amplitude $\varepsilon = 1$.

Figure 2.10 displays the calculated profiles of the Lyman-beta line for the scaled laser amplitude $\varepsilon = 1$ for four different observation angles: $\pi/2$ (solid line), $\pi/3$ (dash-dotted line), $\pi/6$ (dotted line), and 0 (dashed line).

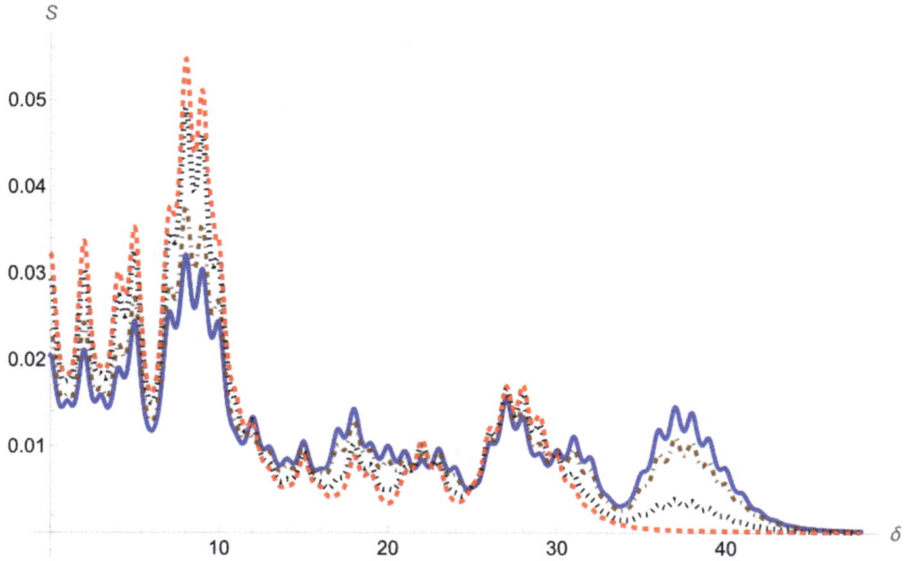

Figure 2.11. Calculated profiles of the Lyman-delta line as in figure 2.7, but for the scaled laser amplitude $\varepsilon = 2$.

Figure 2.11 shows the calculated profiles of the Lyman-beta line for the scaled laser amplitude $\varepsilon = 2$ for four different observation angles: $\pi/2$ (solid line), $\pi/3$ (dash-dotted line), $\pi/6$ (dotted line), and 0 (dashed line).

Figure 2.12 demonstrates the calculated profiles of the Lyman-beta line for the scaled laser amplitude $\varepsilon = 4$ for four different observation angles: $\pi/2$ (solid line), $\pi/3$ (dash-dotted line), $\pi/6$ (dotted line), and 0 (dashed line).

From figures 2.7–2.12 one can see the following. First, as the observation angle θ decreases from $\pi/2$ to 0, the profile narrows: its full width at half maximum decreases. This effect is more pronounced for relatively small values of the scaled laser amplitude ε—just as for the Lyman-beta line.

Second, as the scaled laser amplitude ε increases, the profiles exhibit more and more structures. At the same time, the primary maximum of the profiles shifts to larger distances from the unperturbed position of the Lyman-delta line. However, in distinction to the Lyman-beta line, there appear significant secondary maxima in the wings.

Now we proceed to studying the angular dependence of the Lyman-7 profiles. Figure 2.13 shows its profiles for the scaled laser amplitude $\varepsilon = 0.0625$ for four different observation angles: $\pi/2$ (solid line), $\pi/3$ (dash-dotted line), $\pi/6$ (dotted line), and 0 (dashed line).

Figure 2.14 demonstrates the calculated profiles of the Lyman-7 line for the scaled laser amplitude $\varepsilon = 0.125$ for four different observation angles: $\pi/2$ (solid line), $\pi/3$ (dash-dotted line), $\pi/6$ (dotted line), and 0 (dashed line).

Figure 2.15 displays the calculated profiles of the Lyman-7 line for the scaled laser amplitude $\varepsilon = 0.25$ for four different observation angles: $\pi/2$ (solid line), $\pi/3$ (dash-dotted line), $\pi/6$ (dotted line), and 0 (dashed line).

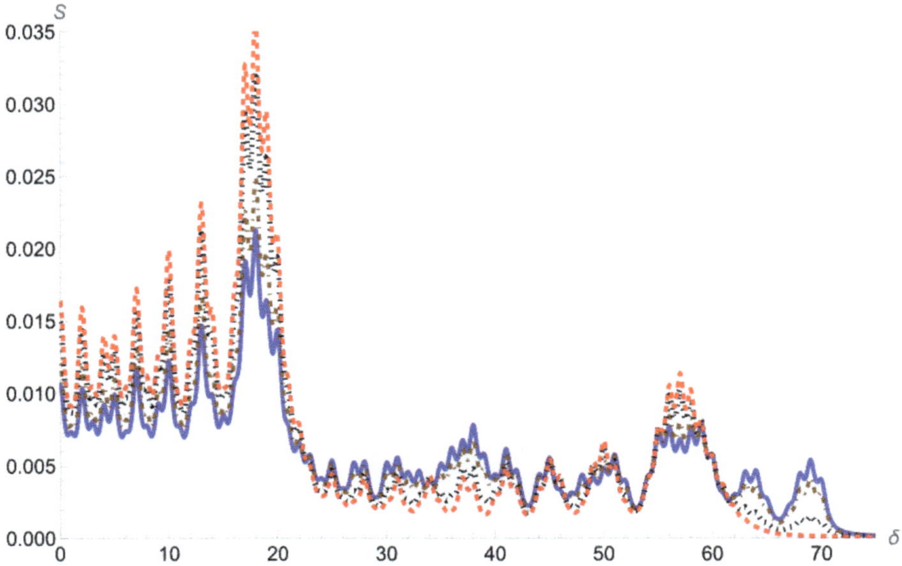

Figure 2.12. Calculated profiles of the Lyman-delta line as in figure 2.7, but for the scaled laser amplitude $\varepsilon = 2$.

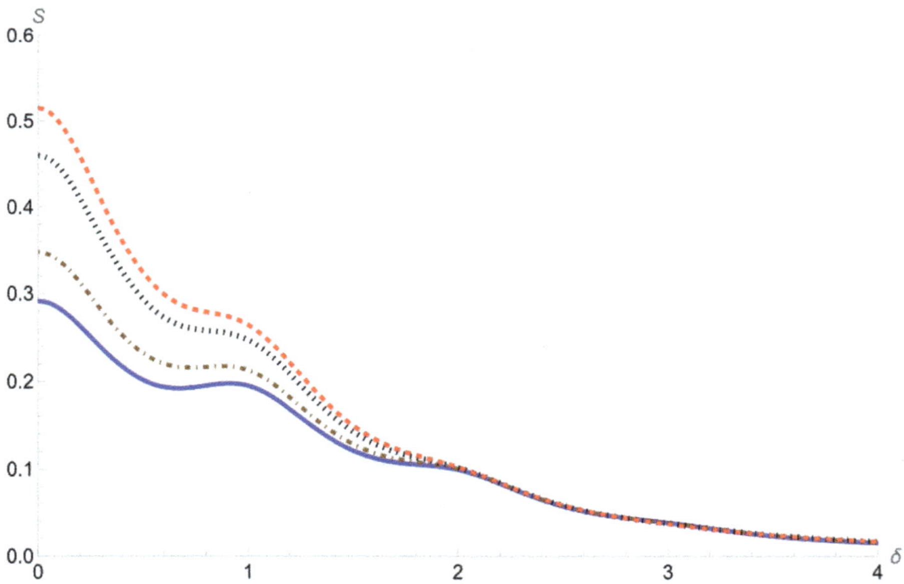

Figure 2.13. Calculated profiles of the Lyman-7 line for the scaled laser amplitude $\varepsilon = 0.0625$ for four different observation angles: $\pi/2$ (solid line), $\pi/3$ (dash-dotted line), $\pi/6$ (dotted line), and 0 (dashed line).

Figure 2.16 shows the calculated profiles of the Lyman-7 line for the scaled laser amplitude $\varepsilon = 0.5$ for four different observation angles: $\pi/2$ (solid line), $\pi/3$ (dash-dotted line), $\pi/6$ (dotted line), and 0 (dashed line).

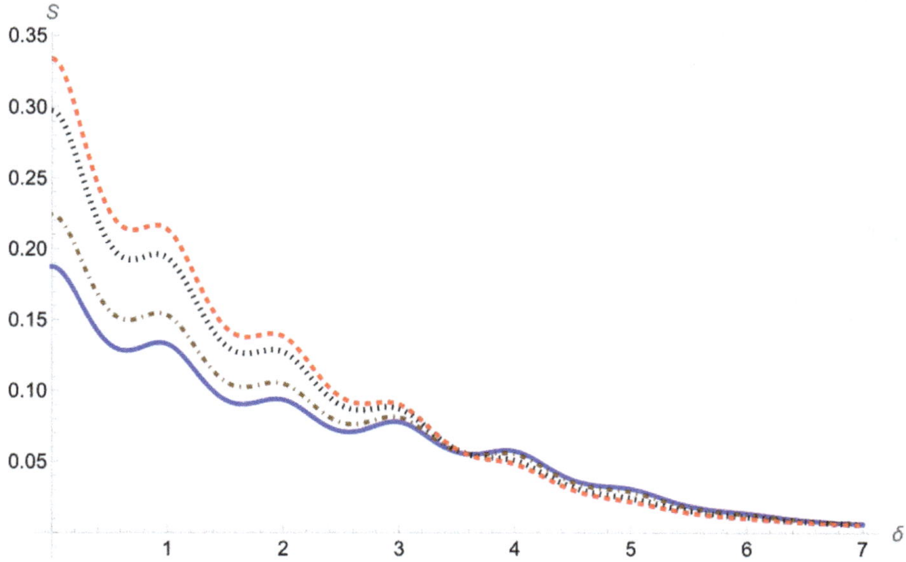

Figure 2.14. Calculated profiles of the Lyman-7 line for the scaled laser amplitude $\varepsilon = 0.125$ for four different observation angles: $\pi/2$ (solid line), $\pi/3$ (dash-dotted line), $\pi/6$ (dotted line), and 0 (dashed line).

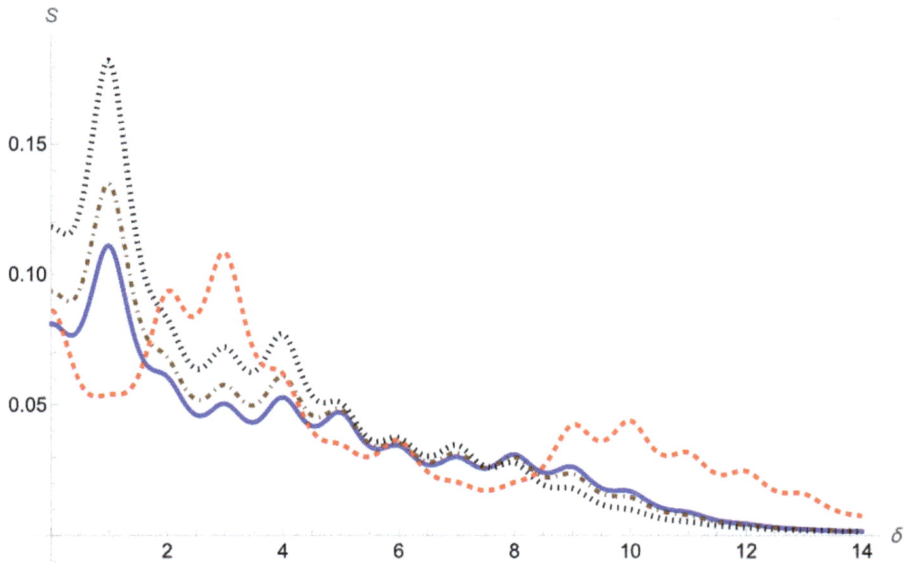

Figure 2.15. Calculated profiles of the Lyman-7 line for the scaled laser amplitude $\varepsilon = 0.25$ for four different observation angles: $\pi/2$ (solid line), $\pi/3$ (dash-dotted line), $\pi/6$ (dotted line), and 0 (dashed line).

Figure 2.17 demonstrates the calculated profiles of the Lyman-7 line for the scaled laser amplitude $\varepsilon = 1$ for four different observation angles: $\pi/2$ (solid line), $\pi/3$ (dash-dotted line), $\pi/6$ (dotted line), and 0 (dashed line).

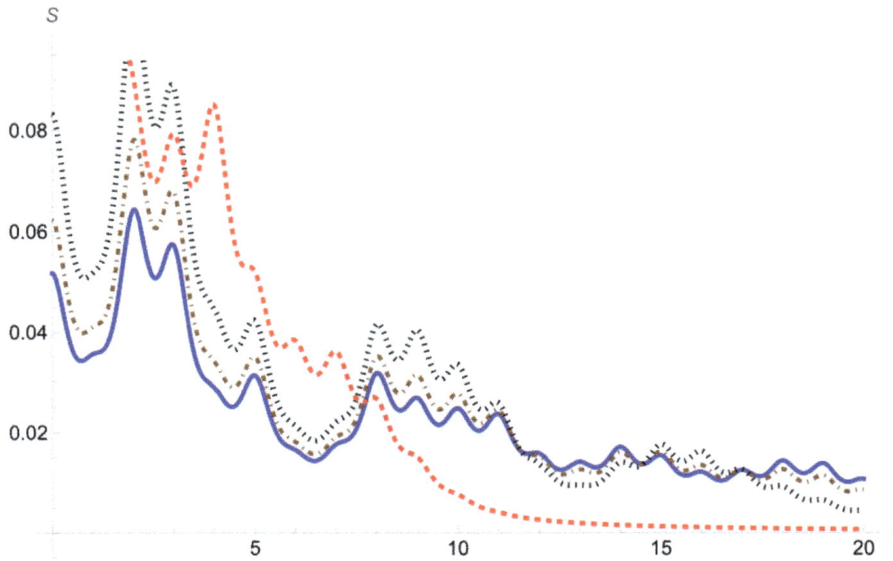

Figure 2.16. Calculated profiles of the Lyman-7 line for the scaled laser amplitude $\varepsilon = 0.5$ for four different observation angles: $\pi/2$ (solid line), $\pi/3$ (dash-dotted line), $\pi/6$ (dotted line), and 0 (dashed line).

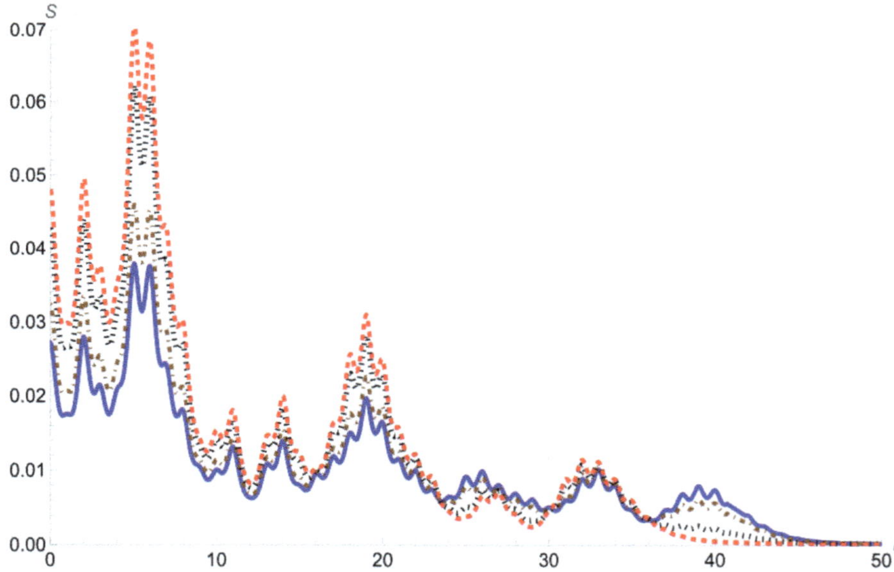

Figure 2.17. Calculated profiles of the Lyman-7 line for the scaled laser amplitude $\varepsilon = 1$ for four different observation angles: $\pi/2$ (solid line), $\pi/3$ (dash-dotted line), $\pi/6$ (dotted line), and 0 (dashed line).

Figure 2.18 displays the calculated profiles of the Lyman-7 line for the scaled laser amplitude $\varepsilon = 2$ for four different observation angles: $\pi/2$ (solid line), $\pi/3$ (dash-dotted line), $\pi/6$ (dotted line), and 0 (dashed line).

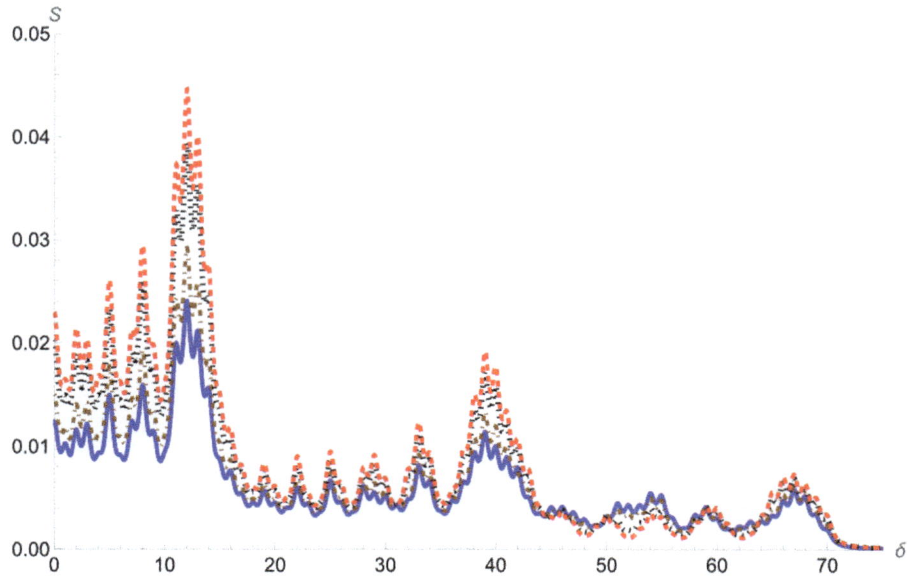

Figure 2.18. Calculated profiles of the Lyman-7 line for the scaled laser amplitude $\varepsilon = 2$ for four different observation angles: $\pi/2$ (solid line), $\pi/3$ (dash-dotted line), $\pi/6$ (dotted line), and 0 (dashed line).

From figures 2.13–2.18 one can see the following. First, as the observation angle θ decreases from $\pi/2$ to 0, the profile narrows: its full width at half maximum decreases. This effect is more pronounced for relatively small values of the scaled laser amplitude ε—just as for the Lyman-beta and Lyman delta lines.

Second, as the scaled laser amplitude ε increases, the profiles exhibit more and more structures. At the same time, the primary maximum of the profiles shifts to larger distances from the unperturbed position of the Lyman-7 line. However, in distinction to the Lyman-beta and Lyman-delta lines, the secondary maxima in the wings become more pronounced.

Now we proceed to studying the angular dependence of the Lyman-9 profiles. Figure 2.19 shows its profiles for the scaled laser amplitude $\varepsilon = 0.031\ 25$ for four different observation angles: $\pi/2$ (solid line), $\pi/3$ (dash-dotted line), $\pi/6$ (dotted line), and 0 (dashed line).

Figure 2.20 demonstrates the calculated profiles of the Lyman-9 line for the scaled laser amplitude $\varepsilon = 0.0625$ for four different observation angles: $\pi/2$ (solid line), $\pi/3$ (dash-dotted line), $\pi/6$ (dotted line), and 0 (dashed line).

Figure 2.21 displays the calculated profiles of the Lyman-9 line for the scaled laser amplitude $\varepsilon = 0.125$ for four different observation angles: $\pi/2$ (solid line), $\pi/3$ (dash-dotted line), $\pi/6$ (dotted line), and 0 (dashed line).

Figure 2.22 shows the calculated profiles of the Lyman-9 line for the scaled laser amplitude $\varepsilon = 0.25$ for four different observation angles: $\pi/2$ (solid line), $\pi/3$ (dash-dotted line), $\pi/6$ (dotted line), and 0 (dashed line).

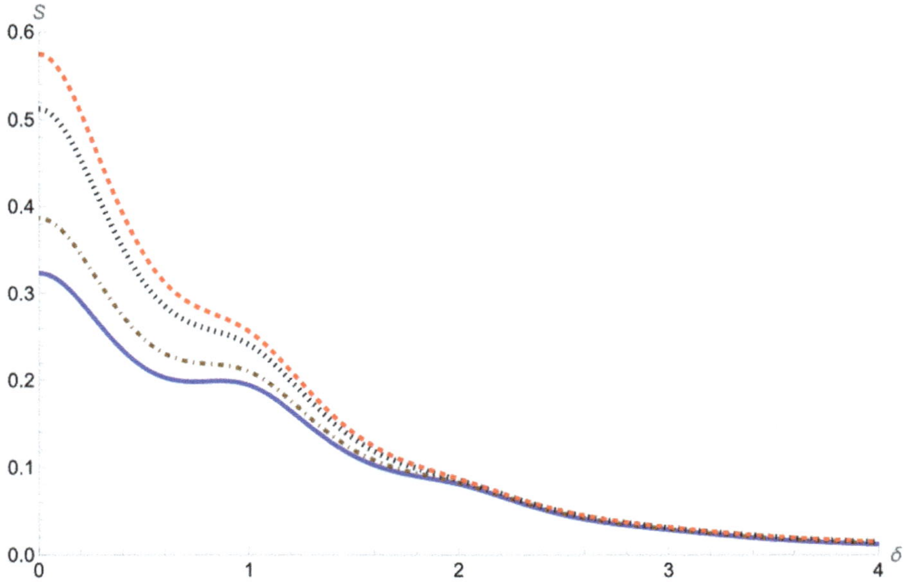

Figure 2.19. Calculated profiles of the Lyman-9 line for the scaled laser amplitude $\varepsilon = 0.031\ 25$ for four different observation angles: $\pi/2$ (solid line), $\pi/3$ (dash-dotted line), $\pi/6$ (dotted line), and 0 (dashed line).

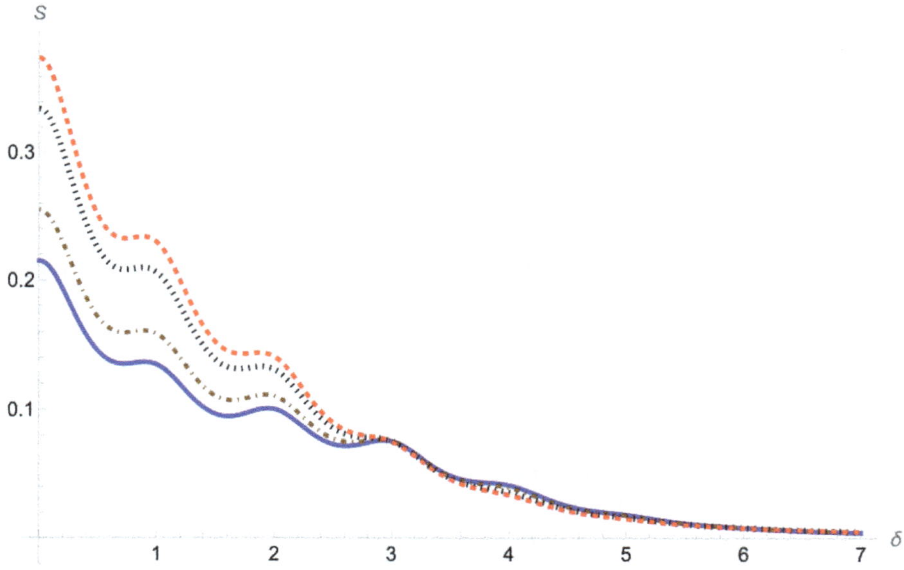

Figure 2.20. Calculated profiles of the Lyman-9 line for the scaled laser amplitude $\varepsilon = 0.0625$ for four different observation angles: $\pi/2$ (solid line), $\pi/3$ (dash-dotted line), $\pi/6$ (dotted line), and 0 (dashed line).

Figure 2.23 demonstrates the calculated profiles of the Lyman-9 line for the scaled laser amplitude $\varepsilon = 0.5$ for four different observation angles: $\pi/2$ (solid line), $\pi/3$ (dash-dotted line), $\pi/6$ (dotted line), and 0 (dashed line).

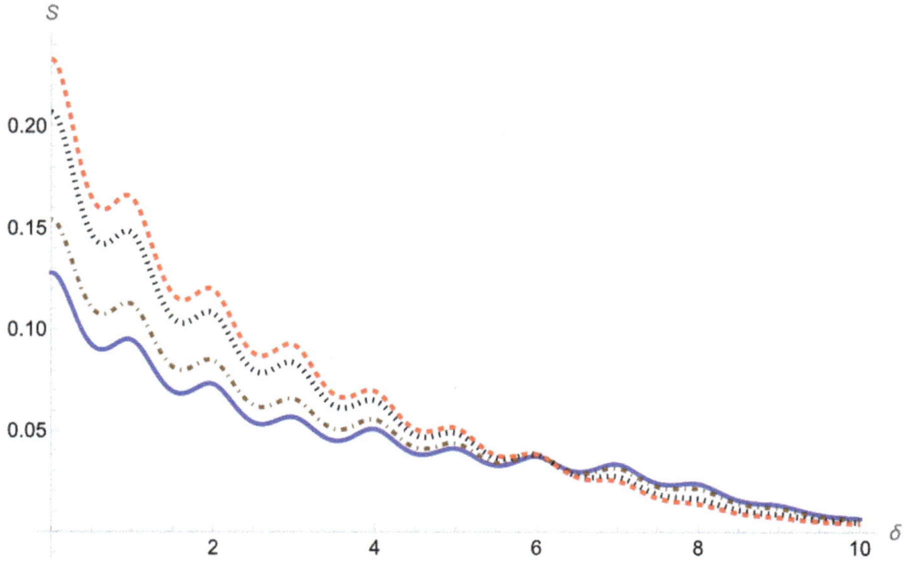

Figure 2.21. Calculated profiles of the Lyman-9 line for the scaled laser amplitude $\varepsilon = 0.125$ for four different observation angles: $\pi/2$ (solid line), $\pi/3$ (dash-dotted line), $\pi/6$ (dotted line), and 0 (dashed line).

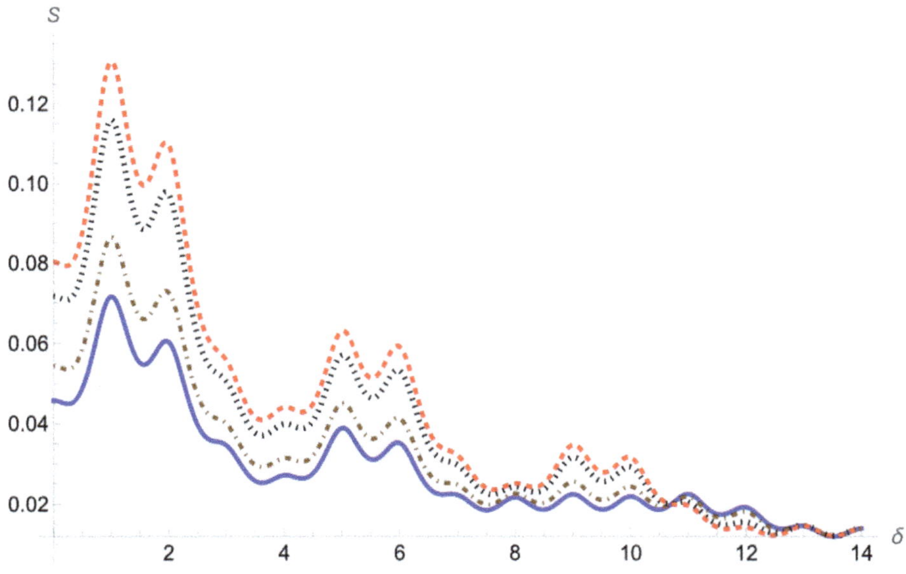

Figure 2.22. Calculated profiles of the Lyman-9 line for the scaled laser amplitude $\varepsilon = 0.25$ for four different observation angles: $\pi/2$ (solid line), $\pi/3$ (dash-dotted line), $\pi/6$ (dotted line), and 0 (dashed line).

Figure 2.24 demonstrates the calculated profiles of the Lyman-9 line for the scaled laser amplitude $\varepsilon = 1$ for four different observation angles: $\pi/2$ (solid line), $\pi/3$ (dash-dotted line), $\pi/6$ (dotted line), and 0 (dashed line).

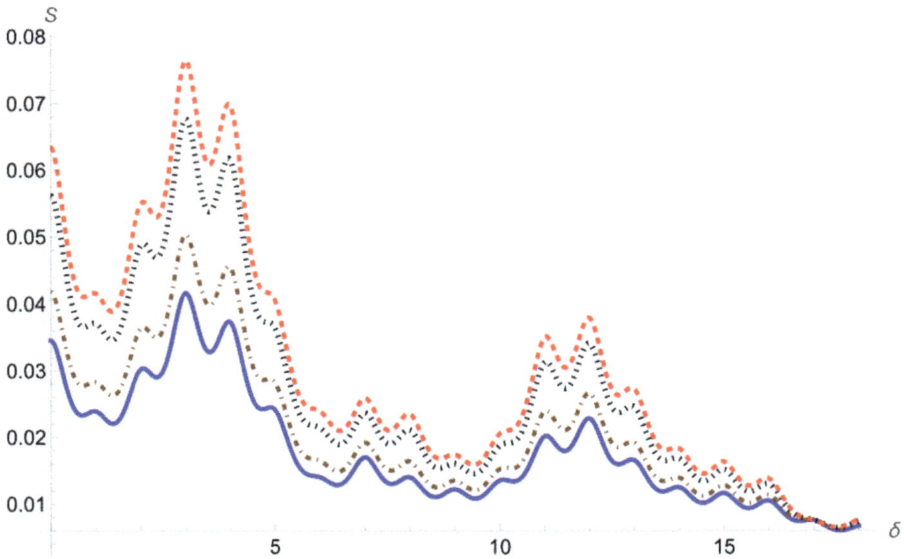

Figure 2.23. Calculated profiles of the Lyman-9 line for the scaled laser amplitude $\varepsilon = 0.5$ for four different observation angles: $\pi/2$ (solid line), $\pi/3$ (dash-dotted line), $\pi/6$ (dotted line), and 0 (dashed line).

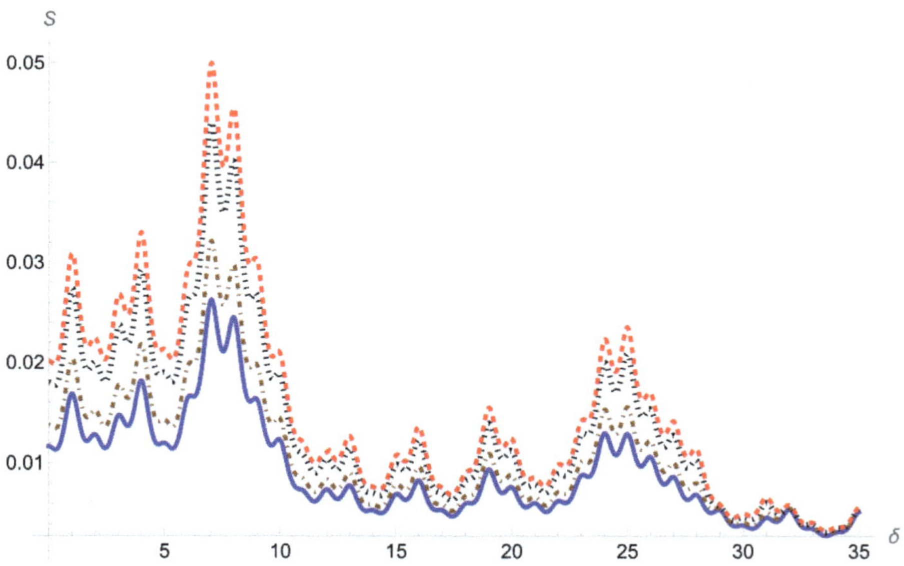

Figure 2.24. Calculated profiles of the Lyman-9 line for the scaled laser amplitude $\varepsilon = 1$ for four different observation angles: $\pi/2$ (solid line), $\pi/3$ (dash-dotted line), $\pi/6$ (dotted line), and 0 (dashed line).

From figures 2.19–2.24 one can see the following. First, as the observation angle θ decreases from $\pi/2$ to 0, the profile narrows: its full width at half maximum decreases. This effect is more pronounced for relatively small values of the scaled laser amplitude ε—just as for the Lyman-beta, Lyman delta, and Lyman-7 lines.

Second, as the scaled laser amplitude ε increases, the profiles exhibit more and more structures. At the same time, the primary maximum of the profiles shifts to larger distances from the unperturbed position of the Lyman-9 line. In distinction to the Lyman-beta, Lyman-delta, and Lyman-7 lines, the secondary maxima in the wings become much more pronounced.

Now we proceed to studying the angular dependence of the Lyman-11 profiles. Figure 2.25 shows its profiles for the scaled laser amplitude $\varepsilon = 0.031\ 25$ for four different observation angles: $\pi/2$ (solid line), $\pi/3$ (dash-dotted line), $\pi/6$ (dotted line), and 0 (dashed line).

Figure 2.26 demonstrates the calculated profiles of the Lyman-11 line for the scaled laser amplitude $\varepsilon = 0.0625$ for four different observation angles: $\pi/2$ (solid line), $\pi/3$ (dash-dotted line), $\pi/6$ (dotted line), and 0 (dashed line).

Figure 2.27 displays the calculated profiles of the Lyman-11 line for the scaled laser amplitude $\varepsilon = 0.125$ for four different observation angles: $\pi/2$ (solid line), $\pi/3$ (dash-dotted line), $\pi/6$ (dotted line), and 0 (dashed line).

Figure 2.28 shows the calculated profiles of the Lyman-11 line for the scaled laser amplitude $\varepsilon = 0.25$ for four different observation angles: $\pi/2$ (solid line), $\pi/3$ (dash-dotted line), $\pi/6$ (dotted line), and 0 (dashed line).

Figure 2.29 demonstrates the calculated profiles of the Lyman-11 line for the scaled laser amplitude $\varepsilon = 0.5$ for four different observation angles: $\pi/2$ (solid line), $\pi/3$ (dash-dotted line), $\pi/6$ (dotted line), and 0 (dashed line).

Figure 2.30 demonstrates the calculated profiles of the Lyman-11 line for the scaled laser amplitude $\varepsilon = 1$ for four different observation angles: $\pi/2$ (solid line), $\pi/3$ (dash-dotted line), $\pi/6$ (dotted line), and 0 (dashed line).

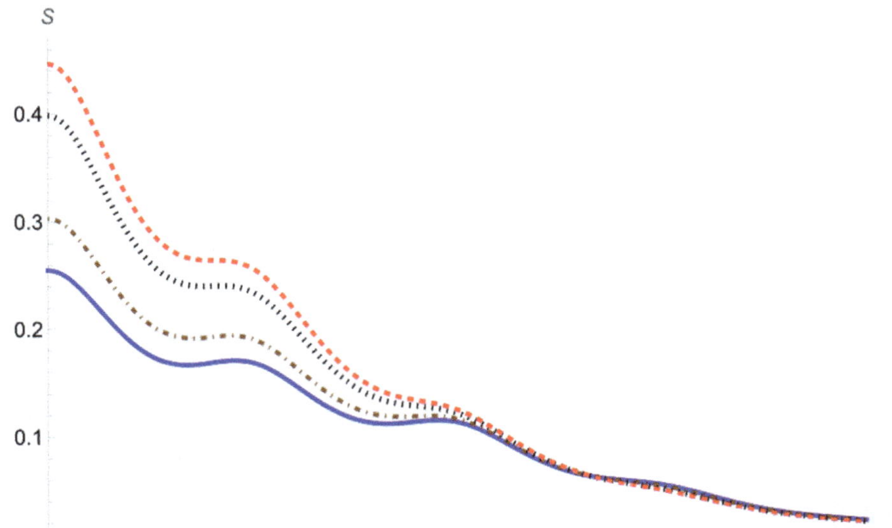

Figure 2.25. Calculated profiles of the Lyman-11 line for the scaled laser amplitude $\varepsilon = 0.031\ 25$ for four different observation angles: $\pi/2$ (solid line), $\pi/3$ (dash-dotted line), $\pi/6$ (dotted line), and 0 (dashed line).

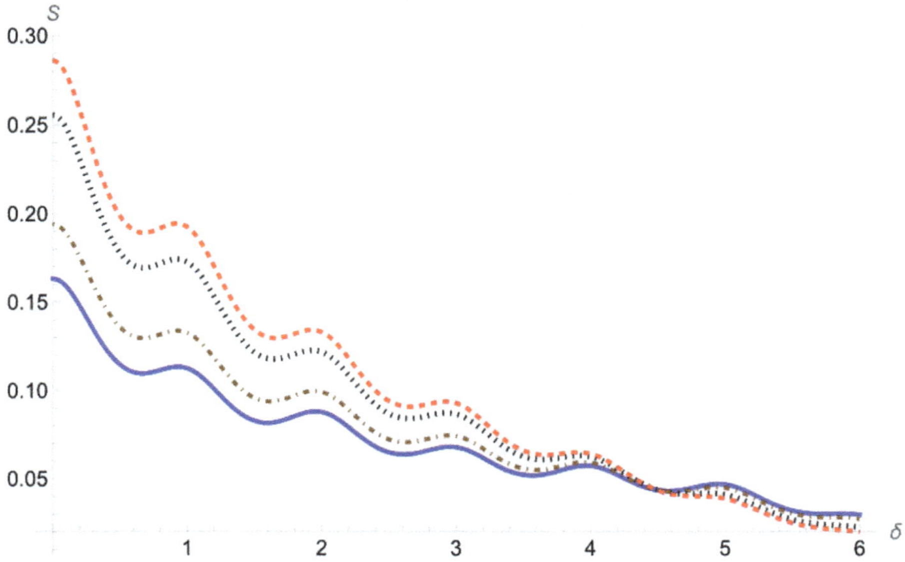

Figure 2.26. Calculated profiles of the Lyman-11 line for the scaled laser amplitude $\varepsilon = 0.0625$ for four different observation angles: $\pi/2$ (solid line), $\pi/3$ (dash-dotted line), $\pi/6$ (dotted line), and 0 (dashed line).

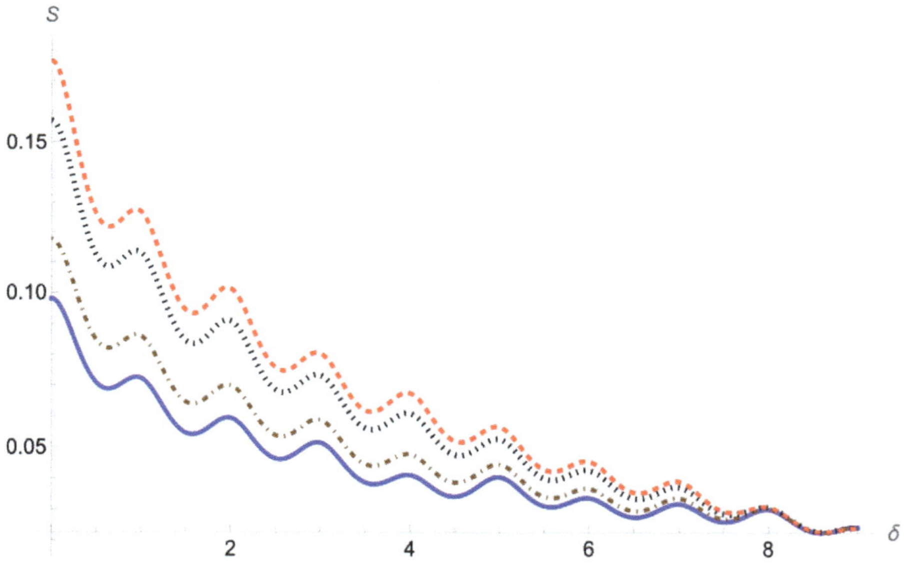

Figure 2.27. Calculated profiles of the Lyman-11 line for the scaled laser amplitude $\varepsilon = 0.125$ for four different observation angles: $\pi/2$ (solid line), $\pi/3$ (dash-dotted line), $\pi/6$ (dotted line), and 0 (dashed line).

From figures 2.25–2.30 one can see the following. First, the profiles of the Lyman-11 line at different observation angles are much more sensitive to the laser field than the corresponding profiles of the Lyman-beta, Lyman-delta, and Lyman-7 lines. Therefore, they can be used for measuring the amplitude of the relatively weak laser field.

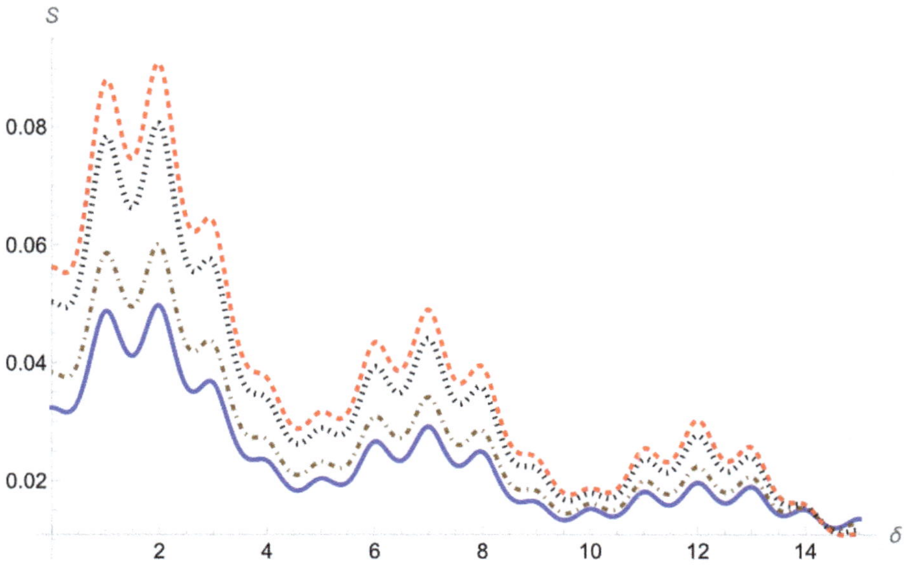

Figure 2.28. Calculated profiles of the Lyman-11 line for the scaled laser amplitude $\varepsilon = 0.25$ for four different observation angles: $\pi/2$ (solid line), $\pi/3$ (dash-dotted line), $\pi/6$ (dotted line), and 0 (dashed line).

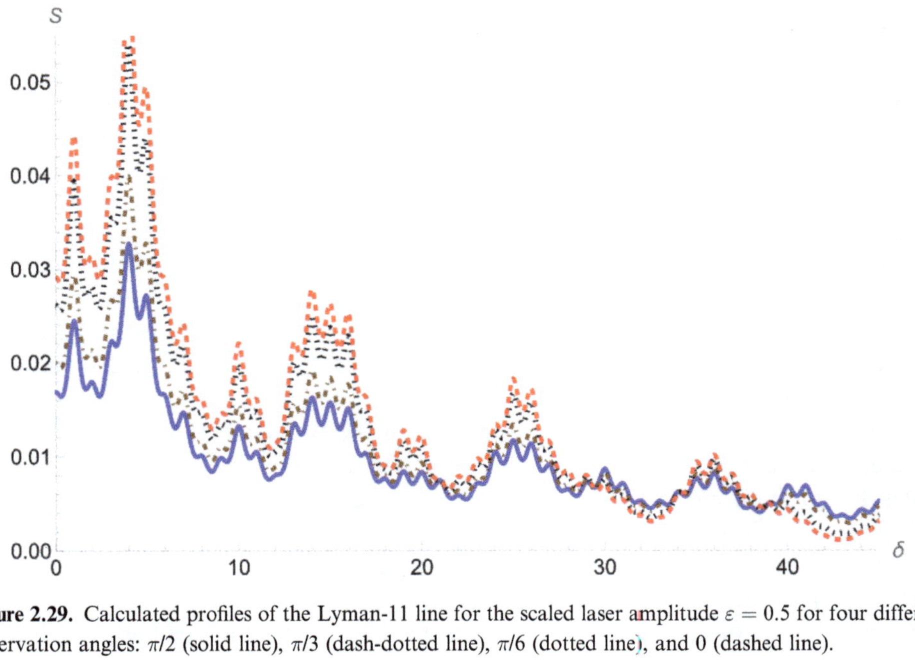

Figure 2.29. Calculated profiles of the Lyman-11 line for the scaled laser amplitude $\varepsilon = 0.5$ for four different observation angles: $\pi/2$ (solid line), $\pi/3$ (dash-dotted line), $\pi/6$ (dotted line), and 0 (dashed line).

Second, as the observation angle θ decreases from $\pi/2$ to 0, the profile narrows: its full width at half maximum decreases. This effect is more pronounced for relatively small values of the scaled laser amplitude ε—just as for the Lyman-beta, Lyman delta, Lyman-7, and Lyman-9 lines.

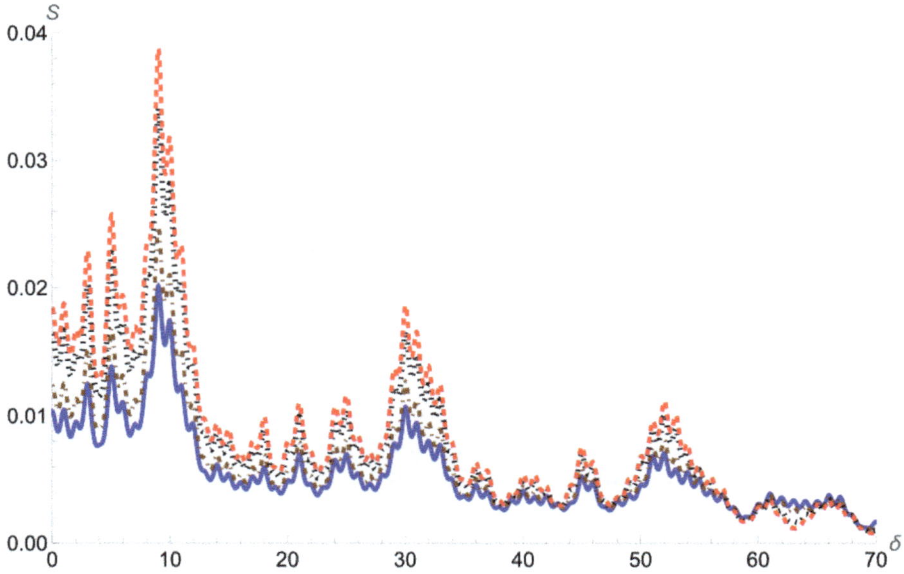

Figure 2.30. Calculated profiles of the Lyman-11 line for the scaled laser amplitude $\varepsilon = 1$ for four different observation angles: $\pi/2$ (solid line), $\pi/3$ (dash-dotted line), $\pi/6$ (dotted line), and 0 (dashed line).

Third, as the scaled laser amplitude ε increases, the profiles exhibit more and more structures. At the same time, the primary maximum of the profiles shifts to larger distances from the unperturbed position of the Lyman-11 line. In distinction to the Lyman-beta, Lyman-delta, and Lyman-7 lines, the secondary maxima in the wings become very significantly pronounced.

Now we proceed to studying the angular dependence of the Balmer-beta profiles. Figure 2.31 shows its profiles for the scaled laser amplitude $\varepsilon = 0.125$ for four different observation angles: $\pi/2$ (solid line), $\pi/3$ (dash-dotted line), $\pi/6$ (dotted line), and 0 (dashed line).

Figure 2.32 demonstrates the calculated profiles of the Balmer-beta line for the scaled laser amplitude $\varepsilon = 0.25$ for four different observation angles: $\pi/2$ (solid line), $\pi/3$ (dash-dotted line), $\pi/6$ (dotted line), and 0 (dashed line).

Figure 2.33 displays the calculated profiles of the Balmer-beta line for the scaled laser amplitude $\varepsilon = 0.5$ for four different observation angles: $\pi/2$ (solid line), $\pi/3$ (dash-dotted line), $\pi/6$ (dotted line), and 0 (dashed line).

Figure 2.34 shows the calculated profiles of the Balmer-beta line for the scaled laser amplitude $\varepsilon = 1$ for four different observation angles: $\pi/2$ (solid line), $\pi/3$ (dash-dotted line), $\pi/6$ (dotted line), and 0 (dashed line).

Figure 2.35 demonstrates the calculated profiles of the Balmer-beta line for the scaled laser amplitude $\varepsilon = 2$ for four different observation angles: $\pi/2$ (solid line), $\pi/3$ (dash-dotted line), $\pi/6$ (dotted line), and 0 (dashed line).

Figure 2.36 demonstrates the calculated profiles of the Balmer-beta line for the scaled laser amplitude $\varepsilon = 4$ for four different observation angles: $\pi/2$ (solid line), $\pi/3$ (dash-dotted line), $\pi/6$ (dotted line), and 0 (dashed line).

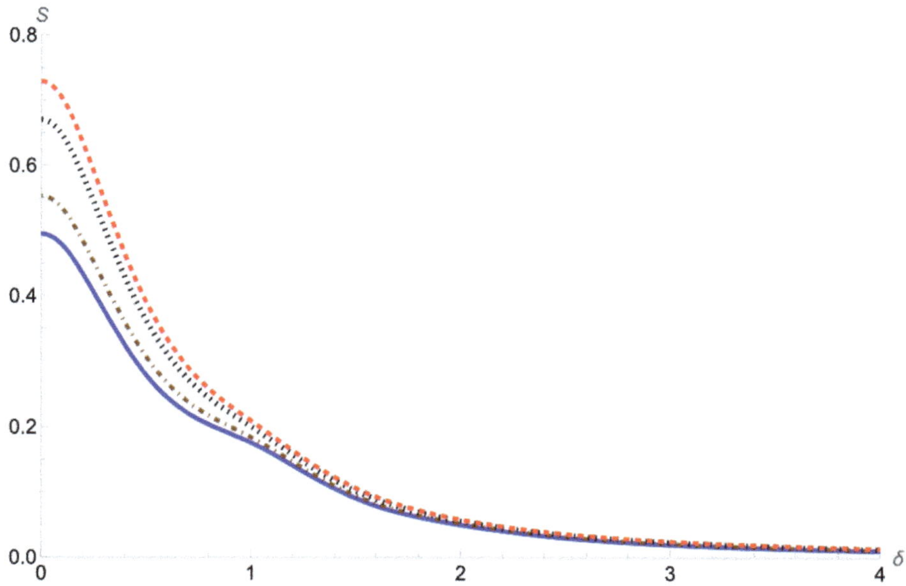

Figure 2.31. Calculated profiles of the Balmer-beta line for the scaled laser amplitude $\varepsilon = 0.125$ for four different observation angles: $\pi/2$ (solid line), $\pi/3$ (dash-dotted line), $\pi/6$ (dotted line), and 0 (dashed line).

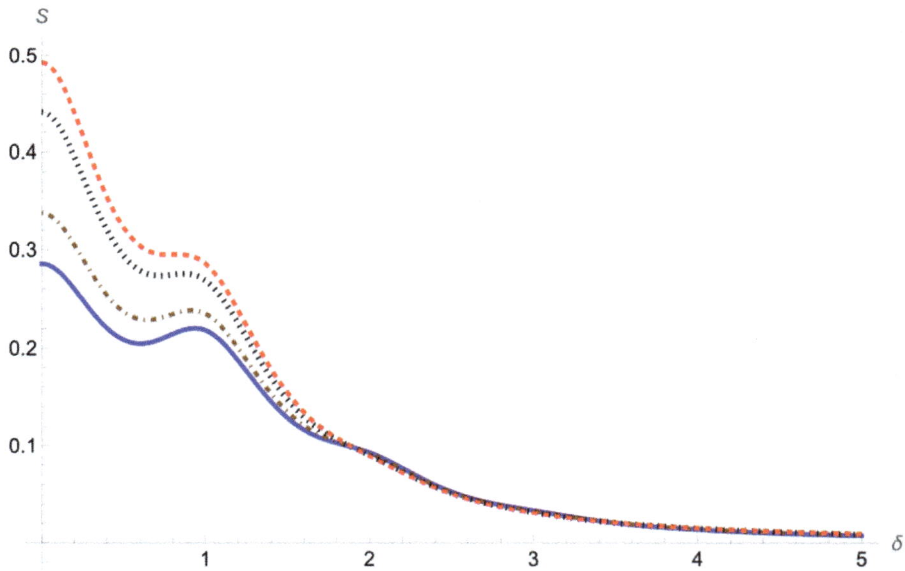

Figure 2.32. Calculated profiles of the Balmer-beta line for the scaled laser amplitude $\varepsilon = 0.25$ for four different observation angles: $\pi/2$ (solid line), $\pi/3$ (dash-dotted line), $\pi/6$ (dotted line), and 0 (dashed line).

From figures 2.31–2.36 one can see the following. First, as the observation angle θ decreases from $\pi/2$ to 0, the profile narrows: its full width at half maximum decreases.

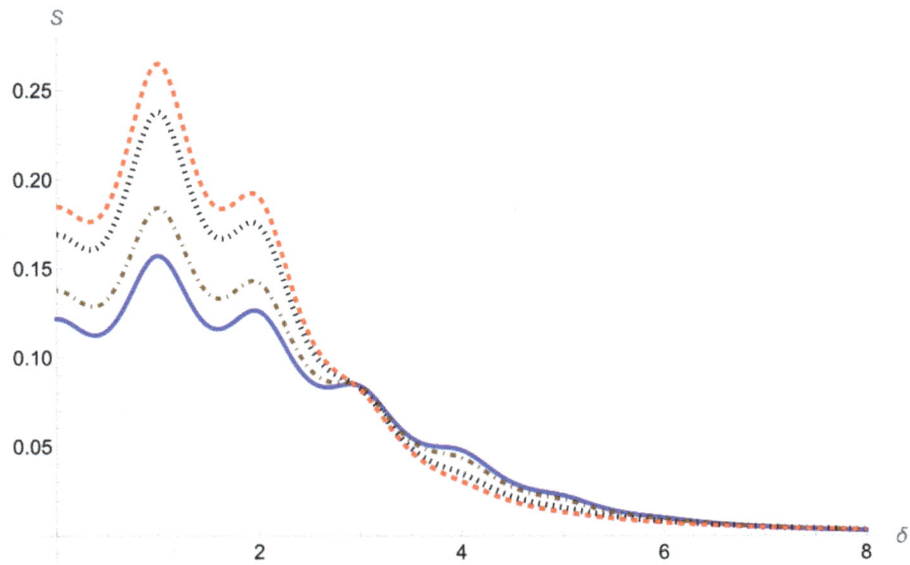

Figure 2.33. Calculated profiles of the Balmer-beta line for the scaled laser amplitude $\varepsilon = 0.5$ for four different observation angles: $\pi/2$ (solid line), $\pi/3$ (dash-dotted line), $\pi/6$ (dotted line), and 0 (dashed line).

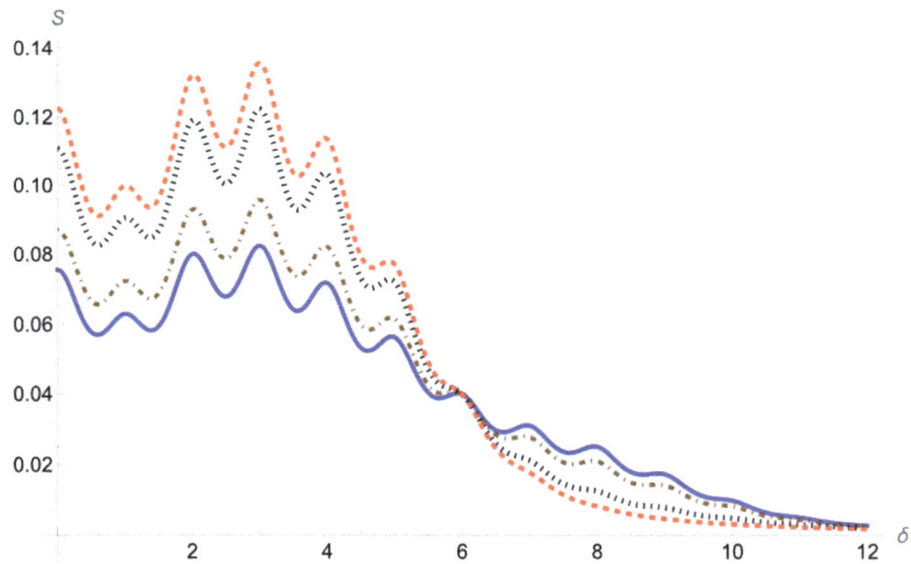

Figure 2.34. Calculated profiles of the Balmer-beta line for the scaled laser amplitude $\varepsilon = 1$ for four different observation angles: $\pi/2$ (solid line), $\pi/3$ (dash-dotted line), $\pi/6$ (dotted line), and 0 (dashed line).

Second, at each observation angle, the profiles are more sensitive to the laser field than the corresponding Lyman-beta and Lyman-delta lines. Therefore, they can be used for measuring the amplitude of a weaker laser field compared to using the Lyman-beta and Lyman-delta lines.

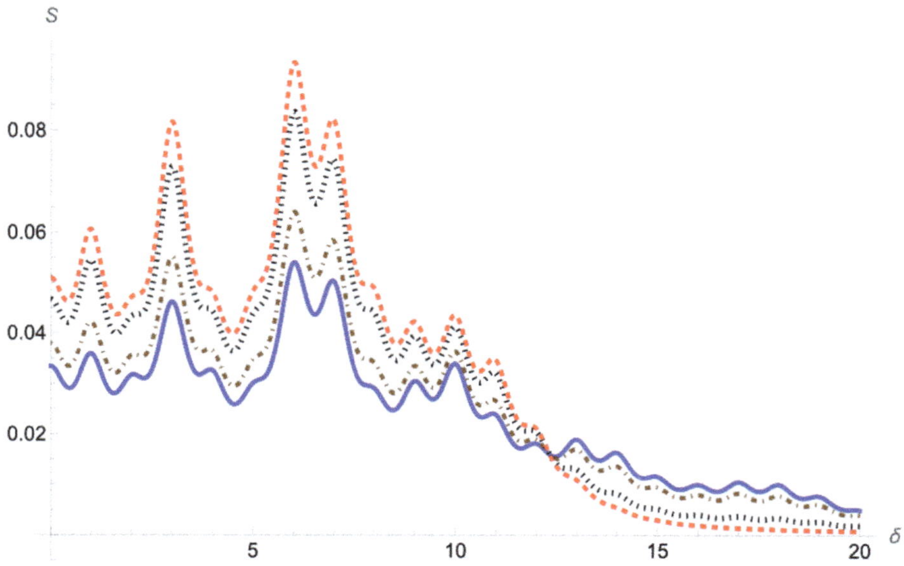

Figure 2.35. Calculated profiles of the Balmer-beta line for the scaled laser amplitude $\varepsilon = 2$ for four different observation angles: $\pi/2$ (solid line), $\pi/3$ (dash-dotted line), $\pi/6$ (dotted line), and 0 (dashed line).

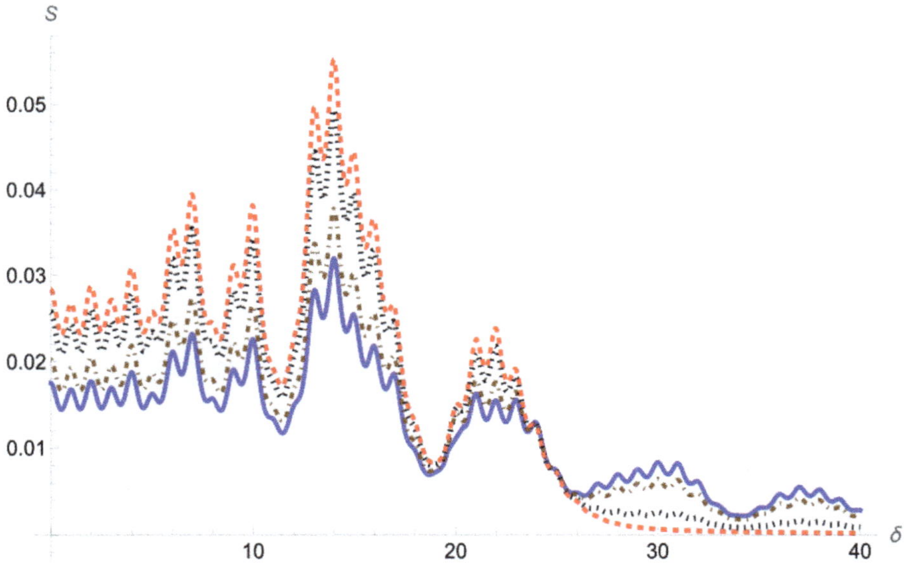

Figure 2.36. Calculated profiles of the Balmer-beta line for the scaled laser amplitude $\varepsilon = 4$ for four different observation angles: $\pi/2$ (solid line), $\pi/3$ (dash-dotted line), $\pi/6$ (dotted line), and 0 (dashed line).

Finally, we proceed to studying the angular dependence of the Balmer-delta profiles. Figure 2.37 shows its profiles for the scaled laser amplitude $\varepsilon = 0.0625$ for four different observation angles: $\pi/2$ (solid line), $\pi/3$ (dash-dotted line), $\pi/6$ (dotted line), and 0 (dashed line).

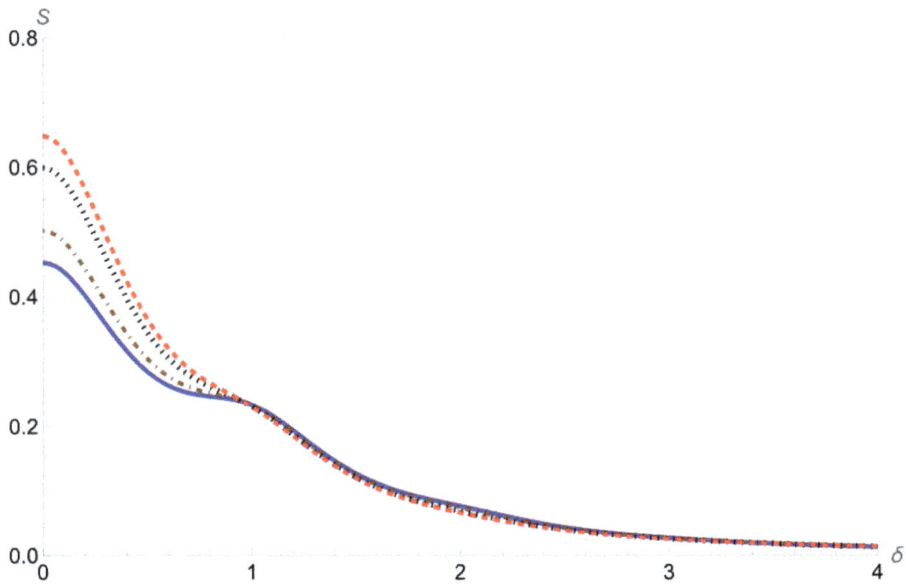

Figure 2.37. Calculated profiles of the Balmer-delta line for the scaled laser amplitude $\varepsilon = 0.0625$ for four different observation angles: $\pi/2$ (solid line), $\pi/3$ (dash-dotted line), $\pi/6$ (dotted line), and 0 (dashed line).

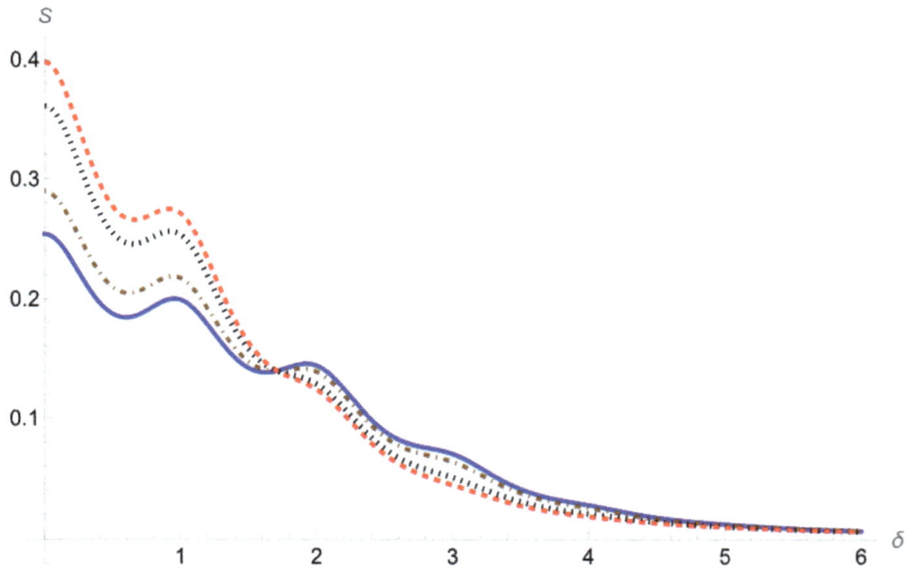

Figure 2.38. Calculated profiles of the Balmer-delta line for the scaled laser amplitude $\varepsilon = 0.125$ for four different observation angles: $\pi/2$ (solid line), $\pi/3$ (dash-dotted line), $\pi/6$ (dotted line), and 0 (dashed line).

Figure 2.38 demonstrates the calculated profiles of the Balmer-delta line for the scaled laser amplitude $\varepsilon = 0.125$ for four different observation angles: $\pi/2$ (solid line), $\pi/3$ (dash-dotted line), $\pi/6$ (dotted line), and 0 (dashed line).

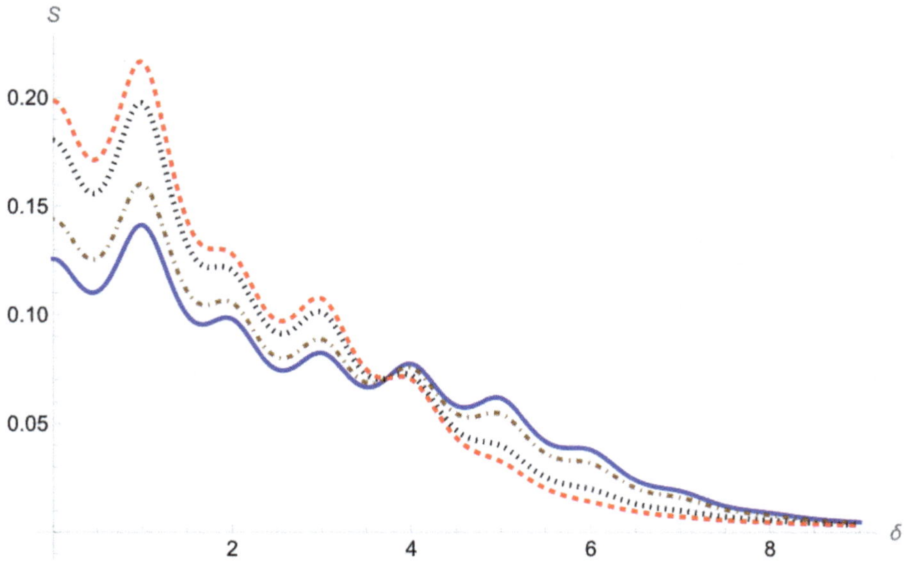

Figure 2.39. Calculated profiles of the Balmer-delta line for the scaled laser amplitude $\varepsilon = 0.25$ for four different observation angles: $\pi/2$ (solid line), $\pi/3$ (dash-dotted line), $\pi/6$ (dotted line), and 0 (dashed line).

Figure 2.39 displays the calculated profiles of the Balmer-delta line for the scaled laser amplitude $\varepsilon = 0.25$ for four different observation angles: $\pi/2$ (solid line), $\pi/3$ (dash-dotted line), $\pi/6$ (dotted line), and 0 (dashed line).

Figure 2.40 shows the calculated profiles of the Balmer-delta line for the scaled laser amplitude $\varepsilon = 0.5$ for four different observation angles: $\pi/2$ (solid line), $\pi/3$ (dash-dotted line), $\pi/6$ (dotted line), and 0 (dashed line).

Figure 2.41 demonstrates the calculated profiles of the Balmer-delta line for the scaled laser amplitude $\varepsilon = 1$ for four different observation angles: $\pi/2$ (solid line), $\pi/3$ (dash-dotted line), $\pi/6$ (dotted line), and 0 (dashed line).

Figure 2.42 displays the calculated profiles of the Balmer-delta line for the scaled laser amplitude $\varepsilon = 2$ for four different observation angles: $\pi/2$ (solid line), $\pi/3$ (dash-dotted line), $\pi/6$ (dotted line), and 0 (dashed line).

From figures 2.37–2.42 one can see the following. First, the profiles of the Balmer-delta line at different observation angles are more sensitive to the laser field than the corresponding profiles of the Balmer-beta line. Therefore, they can be used for measuring the amplitude of the relatively weak laser field.

Second, as the observation angle θ decreases from $\pi/2$ to 0, the profile significantly narrows. Its full width at half maximum decreases.

Third, as the scaled laser amplitude ε increases, the profiles exhibit more and more structures. At the same time, the primary maximum of the profiles shifts to larger distances from the unperturbed position of the Balmer-delta line. The secondary maxima in the wings become very significantly pronounced, just as for the corresponding profiles of the Balmer-beta lines.

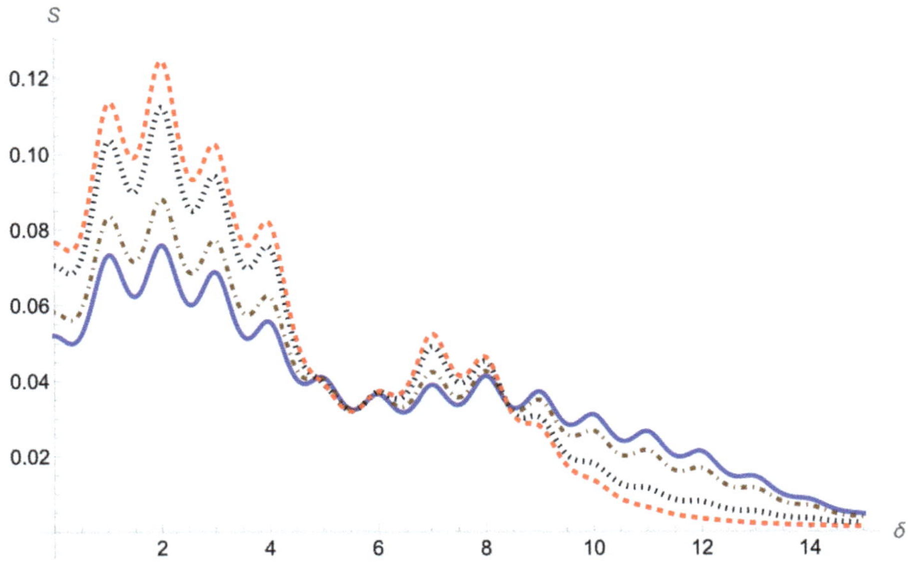

Figure 2.40. Calculated profiles of the Balmer-delta line for the scaled laser amplitude $\varepsilon = 0.5$ for four different observation angles: $\pi/2$ (solid line), $\pi/3$ (dash-dotted line), $\pi/6$ (dotted line), and 0 (dashed line).

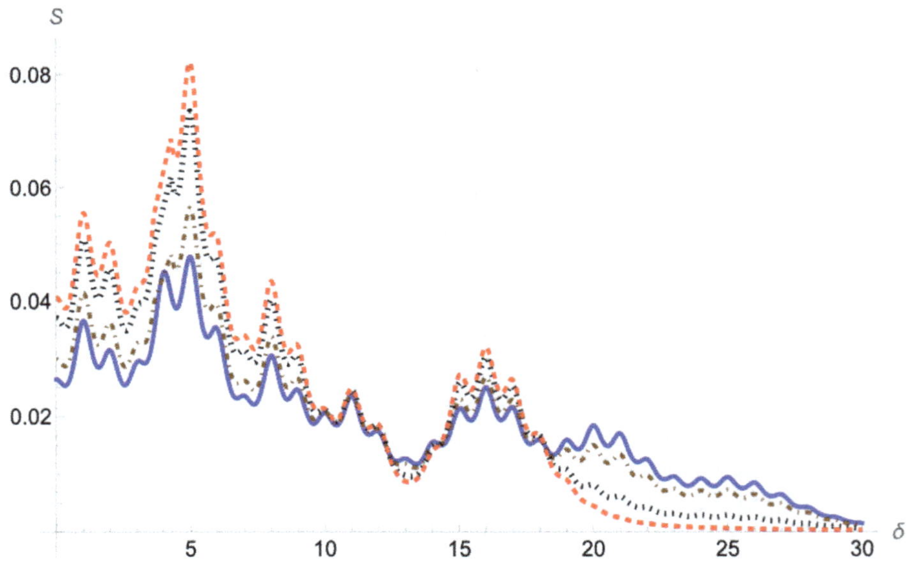

Figure 2.41. Calculated profiles of the Balmer-delta line for the scaled laser amplitude $\varepsilon = 1$ for four different observation angles: $\pi/2$ (solid line), $\pi/3$ (dash-dotted line), $\pi/6$ (dotted line), and 0 (dashed line).

Summarizing section 2.1, we could provide the following application of the results for diagnosing the laser field in plasmas, where due to nonlinear processes, the amplitude of the electromagnetic field at the laser frequency differs from the amplitude of the incoming laser field. The one-dimensional one-mode monochromatic electric

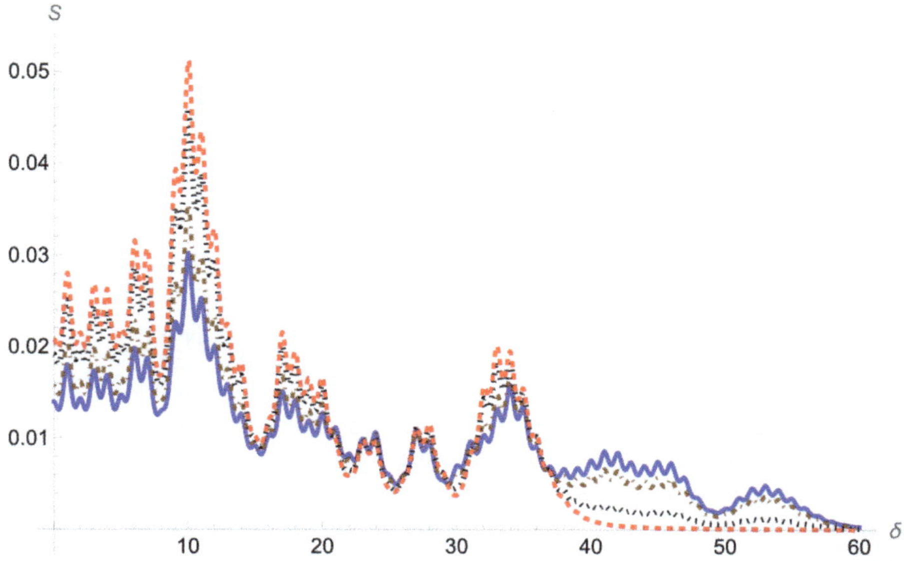

Figure 2.42. Calculated profiles of the Balmer-delta line for the scaled laser amplitude $\varepsilon = 2$ for four different observation angles: $\pi/2$ (solid line), $\pi/3$ (dash-dotted line), $\pi/6$ (dotted line), and 0 (dashed line).

field can create numerous satellites of hydrogenic spectral lines. By trying several hydrogenic lines, an experimentalist can find the line most sensitive to the relevant range of the laser field in the particular experiments. Since, the observed profiles can exhibit numerous satellites, most of the attention should be given to the location and the intensity of the primary maximum and of the secondary maximum. For enhancing the reliability of the determination of the amplitude of the electromagnetic field at the laser frequency inside the plasma, it is a good idea to compare the profiles of the same spectral line observed at different angles—thus taking the advantage of the directional effects described above.

As an example of the experimental observation of Blochizew's type satellites in hydrogenic spectral lines we refer the reader to paper [3].

2.2 Satellites under the one-dimensional two-mode monochromatic electric field

In this case, the profile of a Stark component of a hydrogenic spectral line is (see book [2]):

$$S_{2,\,\text{profile}}(\Delta\omega/\omega) = \sum_{p=-\infty}^{\infty} g(\varepsilon, p)\left(\Delta\omega/\omega - p\right). \tag{2.6}$$

In equation (2.6),

$$g(\varepsilon, p) = J_p^{\,2}(\varepsilon) - J_{p-1}(\varepsilon)J_{p+1}(\varepsilon), \tag{2.7}$$

and

$$\varepsilon = 3\hbar E_{0,\,av}/(2Zm_e e\omega), \qquad E_{0,\,av} = \left(E_1{}^2 + E_2{}^2\right)^{1/2}, \tag{2.8}$$

where E_1 and E_2 are the amplitudes of modes 1 and 2, respectively.

The profile of a multicomponent hydrogenic spectral line can be written in the form:

$$S(\Delta\omega/\omega) = \sum_{p=-\infty}^{+\infty} I(p,\,\varepsilon)\delta(\Delta\omega/\omega) - p), \tag{2.9}$$

$$I(p,\,\varepsilon) = \left[f_0(\theta)\delta_{p0} + 2\sum_{k=1}^{k_{max}} f_k(\theta)g(\varepsilon,\,p)\right]/(f_0 + 2\Sigma f_k).$$

Below we illustrate the angular dependence of profiles of the hydrogenic spectral lines, calculated by equation (2.9), that are the most useful for spectroscopic diagnostics of plasmas. For obtaining continuous profiles we assigned to each satellite the Lorentzian shape of the half width at half maximum equal to $\omega/4$, as in section 2.1.

We start from the Lyman-beta line. Figure 2.43 shows its profile for the scaled laser amplitude $\varepsilon = 0.5$ for four different observation angles: $\pi/2$ (solid line), $\pi/3$ (dash-dotted line), $\pi/6$ (dotted line), and 0 (dashed line).

Figure 2.44 demonstrates the calculated profiles of the Lyman-beta line for the scaled laser amplitude $\varepsilon = 1$ for four different observation angles: $\pi/2$ (solid line), $\pi/3$ (dash-dotted line), $\pi/6$ (dotted line), and 0 (dashed line).

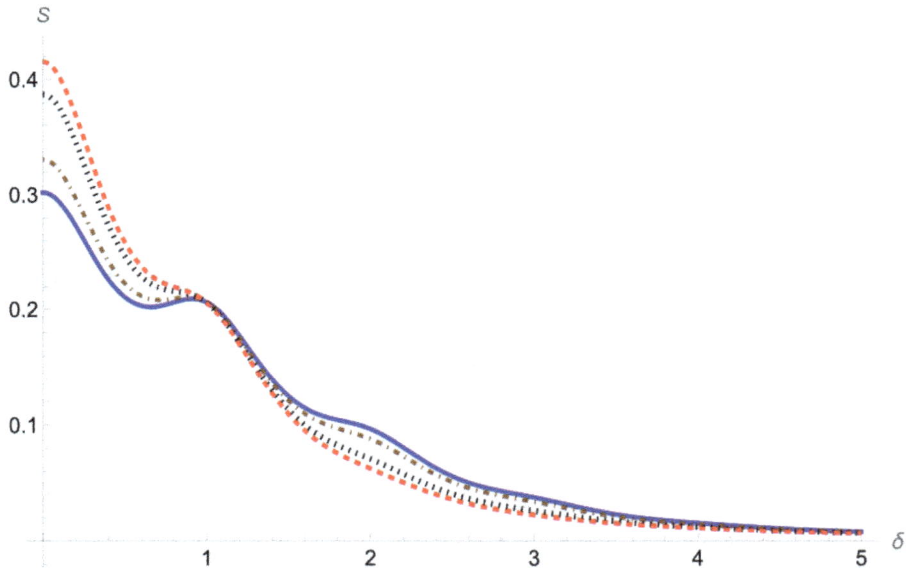

Figure 2.43. Calculated profiles of the Lyman-beta line for the scaled laser amplitude $\varepsilon = 0.5$ for four different observation angles: $\pi/2$ (solid line), $\pi/3$ (dash-dotted line), $\pi/6$ (dotted line), and 0 (dashed line).

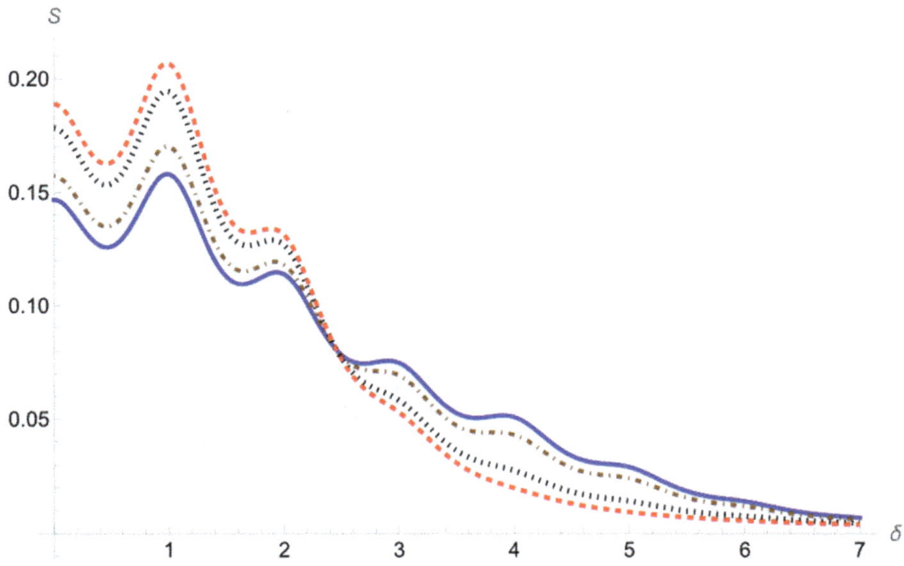

Figure 2.44. Calculated profiles of the Lyman-beta line for the scaled laser amplitude $\varepsilon = 1$ for four different observation angles: $\pi/2$ (solid line), $\pi/3$ (dash-dotted line), $\pi/6$ (dotted line), and 0 (dashed line).

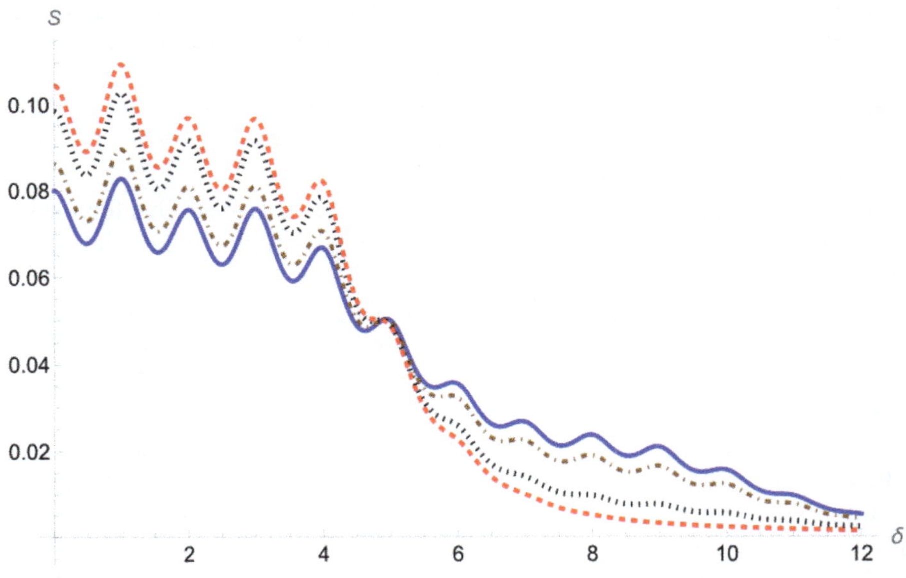

Figure 2.45. Calculated profiles of the Lyman-beta line for the scaled laser amplitude $\varepsilon = 2$ for four different observation angles: $\pi/2$ (solid line), $\pi/3$ (dash-dotted line), $\pi/6$ (dotted line), and 0 (dashed line).

Figure 2.45 displays the calculated profiles of the Lyman-beta line for the scaled laser amplitude $\varepsilon = 2$ for four different observation angles: $\pi/2$ (solid line), $\pi/3$ (dash-dotted line), $\pi/6$ (dotted line), and 0 (dashed line).

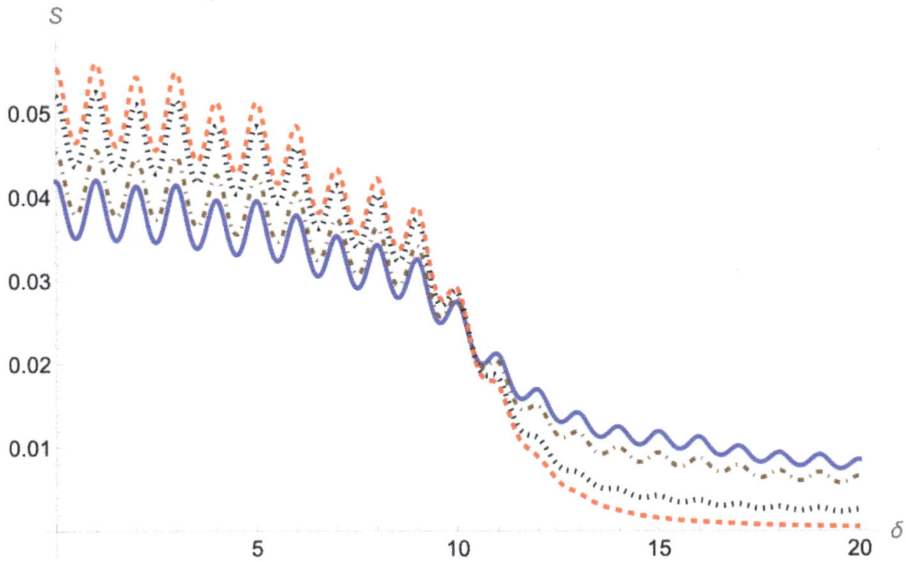

Figure 2.46. Calculated profiles of the Lyman-beta line for the scaled laser amplitude $\varepsilon = 4$ for four different observation angles: $\pi/2$ (solid line), $\pi/3$ (dash-dotted line), $\pi/6$ (dotted line), and 0 (dashed line).

Figure 2.46 shows the calculated profiles of the Lyman-beta line for the scaled laser amplitude $\varepsilon = 4$ for four different observation angles: $\pi/2$ (solid line), $\pi/3$ (dash-dotted line), $\pi/6$ (dotted line), and 0 (dashed line).

Figure 2.47 demonstrates the calculated profiles of the Lyman-beta line for the scaled laser amplitude $\varepsilon = 8$ for four different observation angles: $\pi/2$ (solid line), $\pi/3$ (dash-dotted line), $\pi/6$ (dotted line), and 0 (dashed line).

Figure 2.48 displays the calculated profiles of the Lyman-beta line for the scaled laser amplitude $\varepsilon = 16$ for four different observation angles: $\pi/2$ (solid line), $\pi/3$ (dash-dotted line), $\pi/6$ (dotted line), and 0 (dashed line).

From figures 2.43–2.48 one can see the following. First, the Lyman-beta profiles are sensitive to the direction of the observation. As the angle of the observation decreases from $\pi/2$ to 0, the profile narrows: its full width at half maximum decreases.

Second, there is a clear distinction from the corresponding Lyman-beta profiles for the case of the one-mode laser field. Namely, for the latter case, as the scaled laser amplitude ε increases, the primary maximum of the satellites intensity moves away from the unperturbed position of the spectral line, and the greater the value of ε, the further away the primary maximum moves. However, in the case of the two-mode laser field, the primary maximum of the satellites intensities remains at the unperturbed position of the spectral line.

Third, the peaks of the satellites intensities have a monotonically decreasing envelope. This was not the case for the one-mode laser field.

Now we proceed to studying the angular dependence of the Lyman-delta profiles. Figure 2.49 shows its profiles for the scaled laser amplitude $\varepsilon = 0.125$ for four

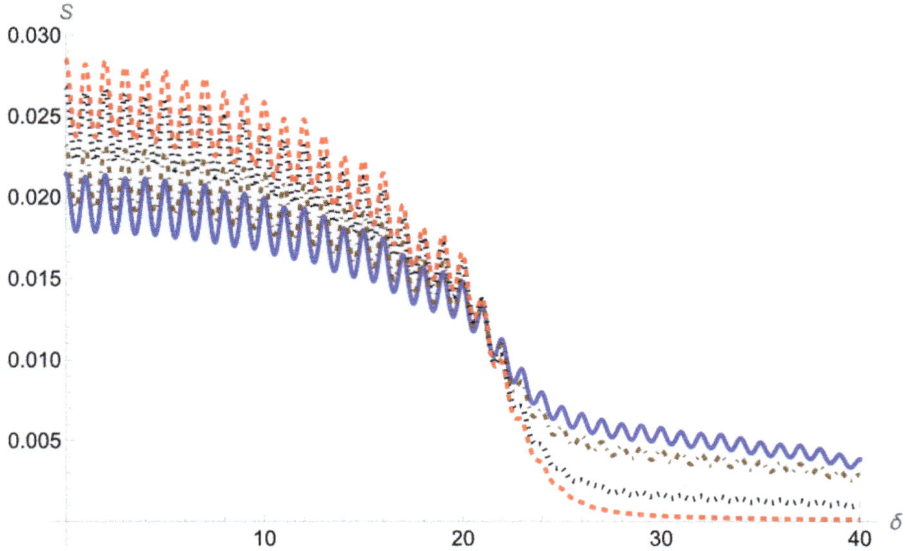

Figure 2.47. Calculated profiles of the Lyman-beta line for the scaled laser amplitude $\varepsilon = 8$ for four different observation angles: $\pi/2$ (solid line), $\pi/3$ (dash-dotted line), $\pi/6$ (dotted line), and 0 (dashed line).

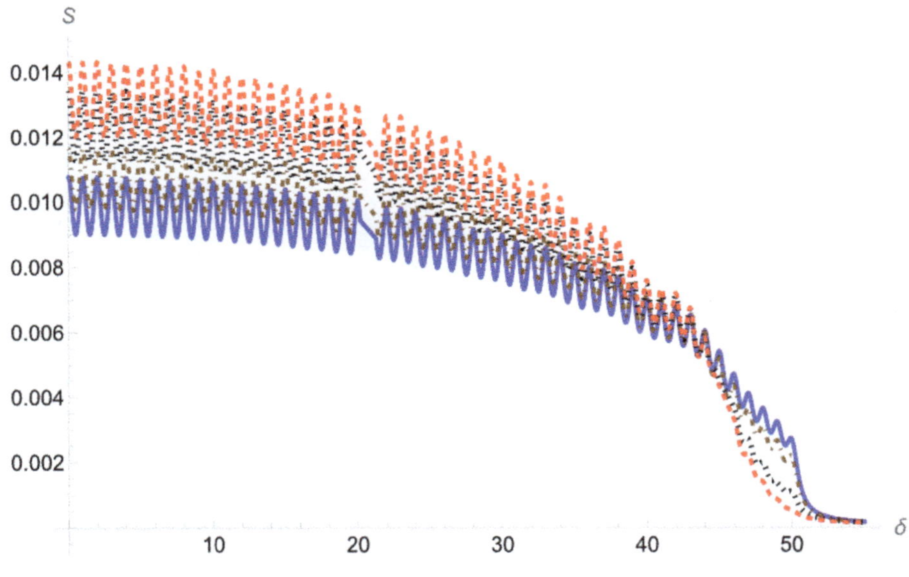

Figure 2.48. Calculated profiles of the Lyman-beta line for the scaled laser amplitude $\varepsilon = 16$ for four different observation angles: $\pi/2$ (solid line), $\pi/3$ (dash-dotted line), $\pi/6$ (dotted line), and 0 (dashed line).

different observation angles: $\pi/2$ (solid line), $\pi/3$ (dash-dotted line), $\pi/6$ (dotted line), and 0 (dashed line).

Figure 2.50 demonstrates the calculated profiles of the Lyman-delta line for the scaled laser amplitude $\varepsilon = 0.25$ for four different observation angles: $\pi/2$ (solid line), $\pi/3$ (dash-dotted line), $\pi/6$ (dotted line), and 0 (dashed line).

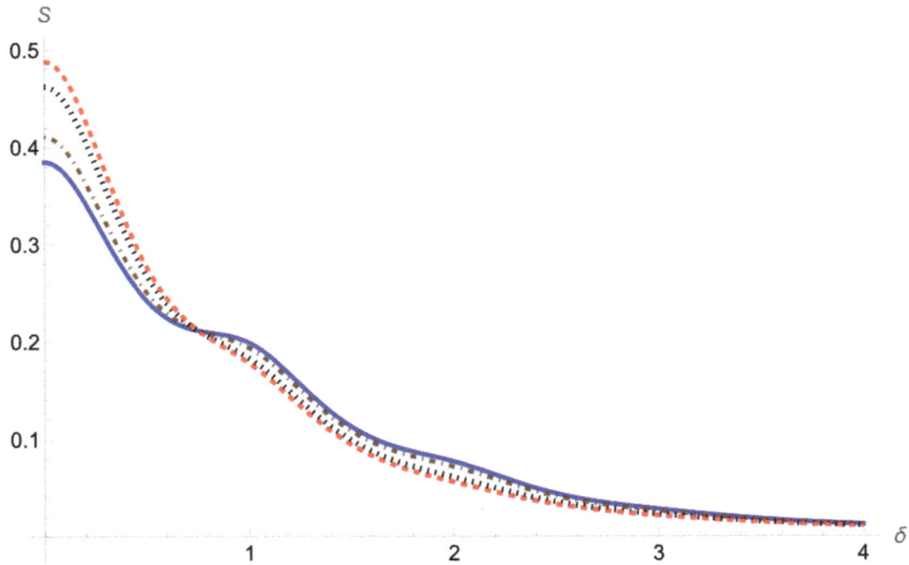

Figure 2.49. Calculated profiles of the Lyman-delta line for the scaled laser amplitude $\varepsilon = 0.125$ for four different observation angles: $\pi/2$ (solid line), $\pi/3$ (dash-dotted line), $\pi/6$ (dotted line), and 0 (dashed line).

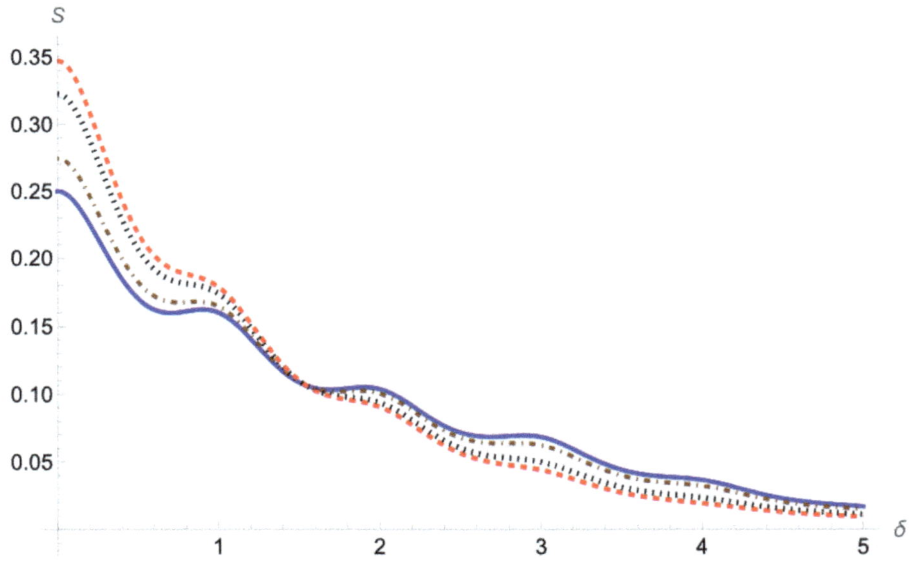

Figure 2.50. Calculated profiles of the Lyman-delta line for the scaled laser amplitude $\varepsilon = 0.25$ for four different observation angles: $\pi/2$ (solid line), $\pi/3$ (dash-dotted line), $\pi/6$ (dotted line), and 0 (dashed line).

Figure 2.51 displays the calculated profiles of the Lyman-delta line for the scaled laser amplitude $\varepsilon = 0.5$ for four different observation angles: $\pi/2$ (solid line), $\pi/3$ (dash-dotted line), $\pi/6$ (dotted line), and 0 (dashed line).

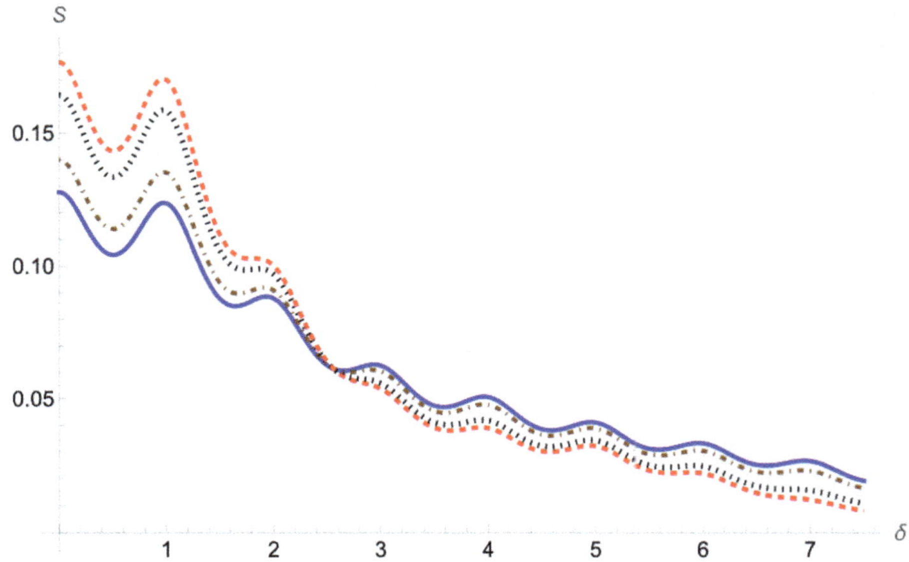

Figure 2.51. Calculated profiles of the Lyman-delta line for the scaled laser amplitude $\varepsilon = 0.5$ for four different observation angles: $\pi/2$ (solid line), $\pi/3$ (dash-dotted line), $\pi/6$ (dotted line), and 0 (dashed line).

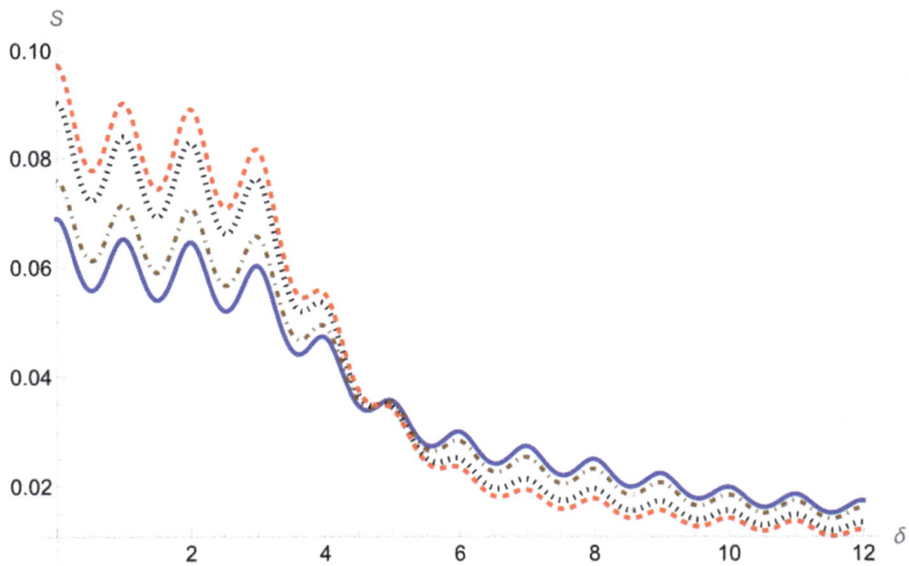

Figure 2.52. Calculated profiles of the Lyman-delta line for the scaled laser amplitude $\varepsilon = 1$ for four different observation angles: $\pi/2$ (solid line), $\pi/3$ (dash-dotted line), $\pi/6$ (dotted line), and 0 (dashed line).

Figure 2.52 shows the calculated profiles of the Lyman-delta line for the scaled laser amplitude $\varepsilon = 1$ for four different observation angles: $\pi/2$ (solid line), $\pi/3$ (dash-dotted line), $\pi/6$ (dotted line), and 0 (dashed line).

Figure 2.53 demonstrates the calculated profiles of the Lyman-delta line for the scaled laser amplitude $\varepsilon = 2$ for four different observation angles: $\pi/2$ (solid line), $\pi/3$ (dash-dotted line), $\pi/6$ (dotted line), and 0 (dashed line).

Figure 2.54 displays the calculated profiles of the Lyman-delta line for the scaled laser amplitude $\varepsilon = 4$ for four different observation angles: $\pi/2$ (solid line), $\pi/3$ (dash-dotted line), $\pi/6$ (dotted line), and 0 (dashed line).

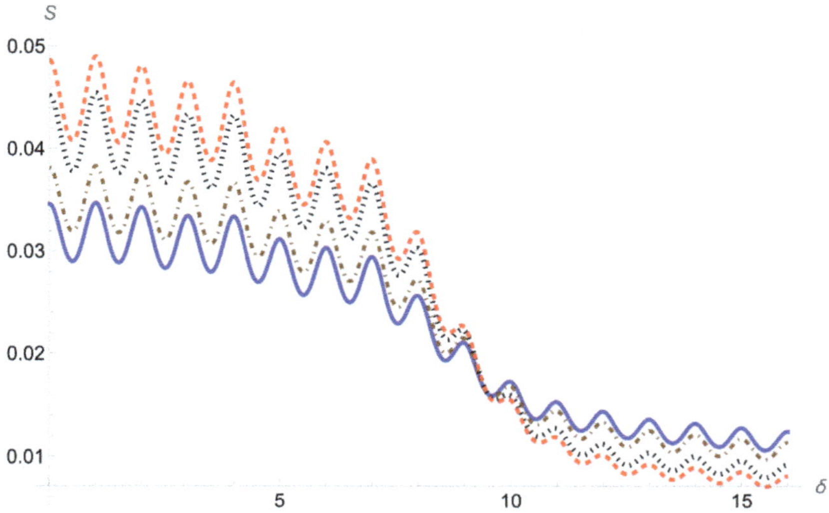

Figure 2.53. Calculated profiles of the Lyman-delta line for the scaled laser amplitude $\varepsilon = 2$ for four different observation angles: $\pi/2$ (solid line), $\pi/3$ (dash-dotted line), $\pi/6$ (dotted line), and 0 (dashed line).

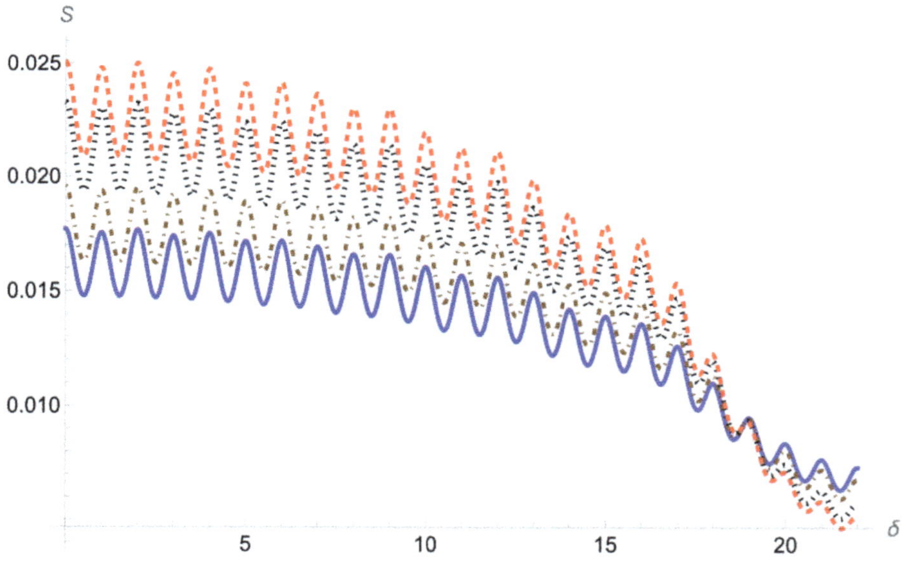

Figure 2.54. Calculated profiles of the Lyman-delta line for the scaled laser amplitude $\varepsilon = 4$ for four different observation angles: $\pi/2$ (solid line), $\pi/3$ (dash-dotted line), $\pi/6$ (dotted line), and 0 (dashed line).

From figures 2.49–2.54 one can see the following. First, the Lyman-delta profiles are more sensitive to the laser field than the Lyman-beta profiles. They are also sensitive to the direction of the observation. As the angle of the observation decreases from $\pi/2$ to 0, the profile narrows.

Second, there is again a clear distinction from the corresponding Lyman-delta profiles for the case of the one-mode laser field. For the latter case, as the scaled laser amplitude ε increases, the primary maximum of the satellites intensity moves away from the unperturbed position of the spectral line, and the greater the value of ε, the further away the primary maximum moves. However, in the case of the two-mode laser field, the primary maximum of the satellites intensities remains at the unperturbed position of the spectral line.

Third, again the peaks of the satellites intensities have a monotonically decreasing envelope. This was not the case for the one-mode laser field.

Now we proceed to studying the angular dependence of the Lyman-7 profiles. Figure 2.55 shows its profiles for the scaled laser amplitude $\varepsilon = 0.0625$ for four different observation angles: $\pi/2$ (solid line), $\pi/3$ (dash-dotted line), $\pi/6$ (dotted line), and 0 (dashed line).

Figure 2.56 demonstrates the calculated profiles of the Lyman-7 line for the scaled laser amplitude $\varepsilon = 0.125$ for four different observation angles: $\pi/2$ (solid line), $\pi/3$ (dash-dotted line), $\pi/6$ (dotted line), and 0 (dashed line).

Figure 2.57 displays the calculated profiles of the Lyman-7 line for the scaled laser amplitude $\varepsilon = 0.25$ for four different observation angles: $\pi/2$ (solid line), $\pi/3$ (dash-dotted line), $\pi/6$ (dotted line), and 0 (dashed line).

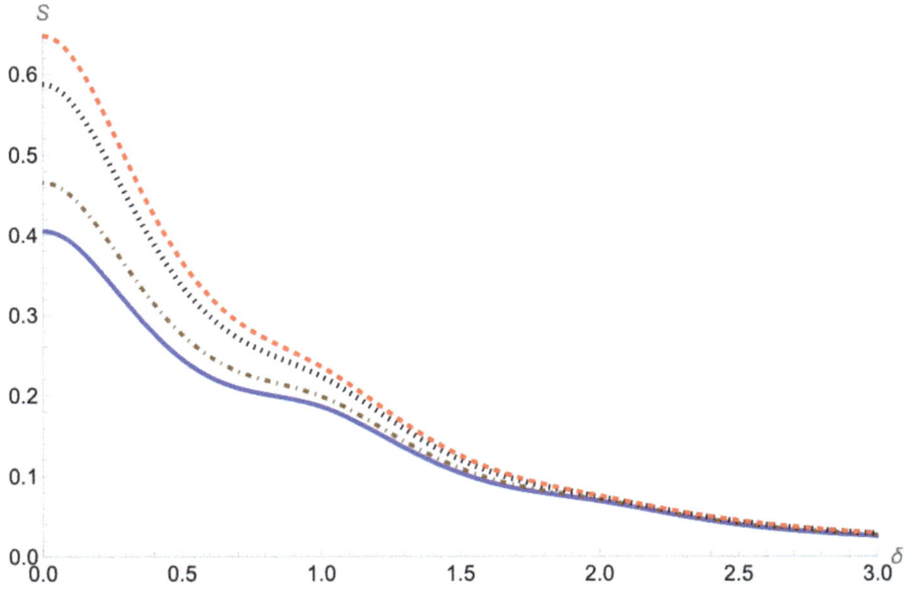

Figure 2.55. Calculated profiles of the Lyman-7 line for the scaled laser amplitude $\varepsilon = 0.0625$ for four different observation angles: $\pi/2$ (solid line), $\pi/3$ (dash-dotted line), $\pi/6$ (dotted line), and 0 (dashed line).

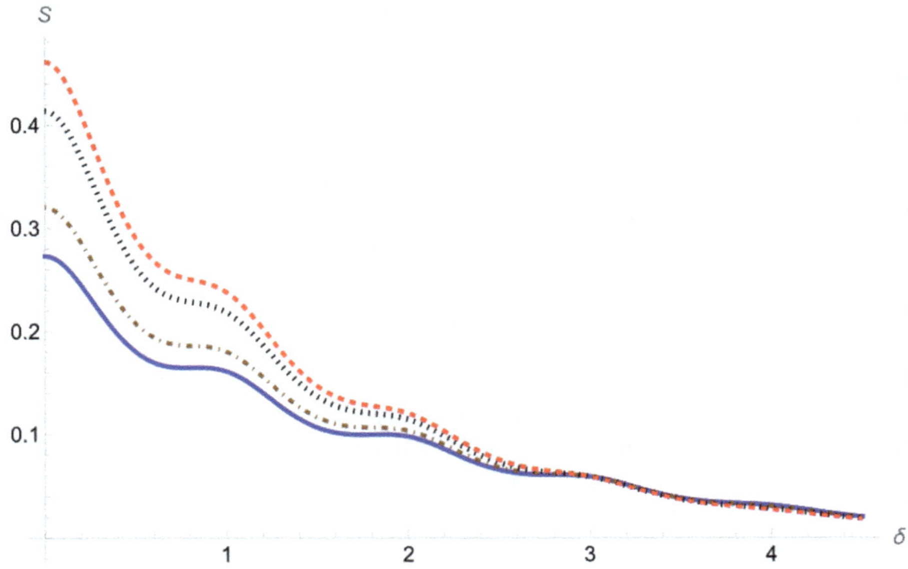

Figure 2.56. Calculated profiles of the Lyman-7 line for the scaled laser amplitude $\varepsilon = 0.125$ for four different observation angles: $\pi/2$ (solid line), $\pi/3$ (dash-dotted line), $\pi/6$ (dotted line), and 0 (dashed line).

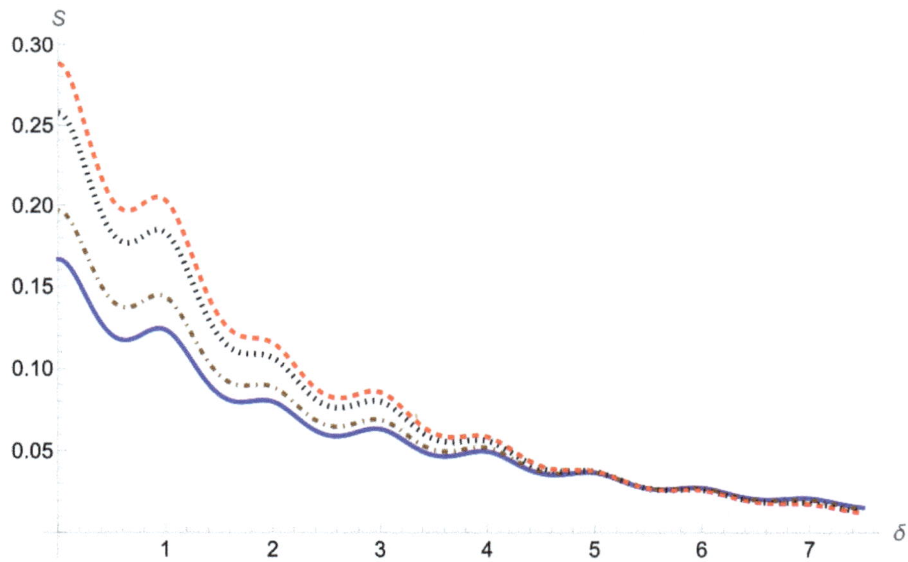

Figure 2.57. Calculated profiles of the Lyman-7 line for the scaled laser amplitude $\varepsilon = 0.25$ for four different observation angles: $\pi/2$ (solid line), $\pi/3$ (dash-dotted line), $\pi/6$ (dotted line), and 0 (dashed line).

Figure 2.58 shows the calculated profiles of the Lyman-7 line for the scaled laser amplitude $\varepsilon = 0.5$ for four different observation angles: $\pi/2$ (solid line), $\pi/3$ (dash-dotted line), $\pi/6$ (dotted line), and 0 (dashed line).

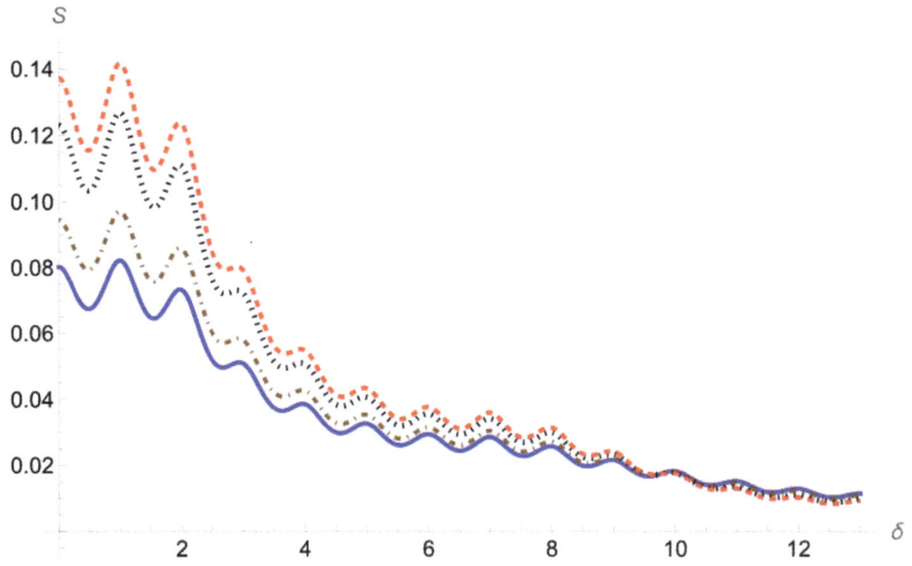

Figure 2.58. Calculated profiles of the Lyman-7 line for the scaled laser amplitude $\varepsilon = 0.5$ for four different observation angles: $\pi/2$ (solid line), $\pi/3$ (dash-dotted line), $\pi/6$ (dotted line), and 0 (dashed line).

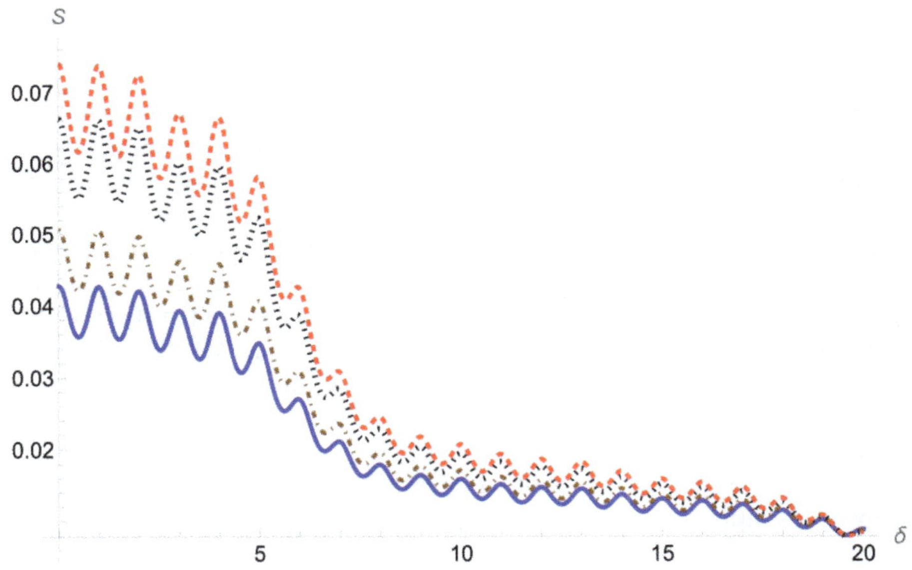

Figure 2.59. Calculated profiles of the Lyman-7 line for the scaled laser amplitude $\varepsilon = 1$ for four different observation angles: $\pi/2$ (solid line), $\pi/3$ (dash-dotted line), $\pi/6$ (dotted line), and 0 (dashed line).

Figure 2.59 demonstrates the calculated profiles of the Lyman-7 line for the scaled laser amplitude $\varepsilon = 1$ for four different observation angles: $\pi/2$ (solid line), $\pi/3$ (dash-dotted line), $\pi/6$ (dotted line), and 0 (dashed line).

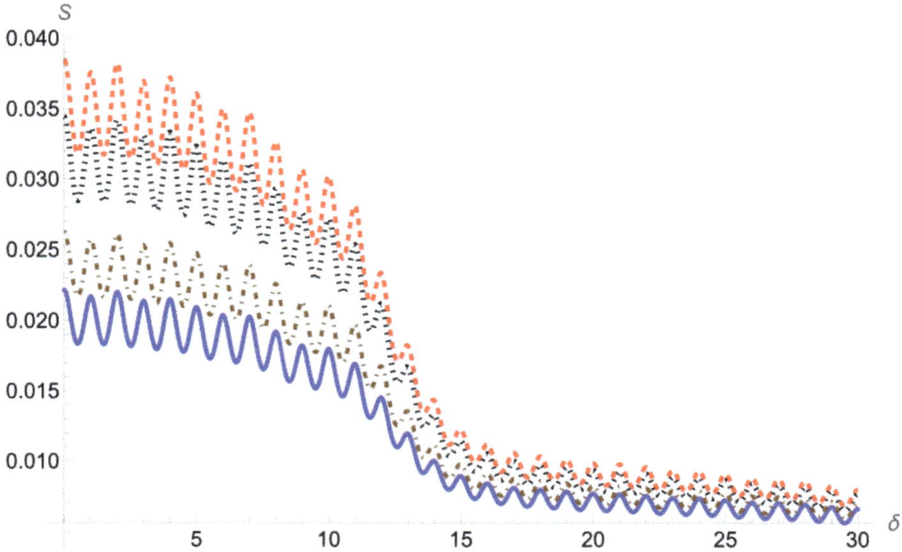

Figure 2.60. Calculated profiles of the Lyman-7 line for the scaled laser amplitude $\varepsilon = 2$ for four different observation angles: $\pi/2$ (solid line), $\pi/3$ (dash-dotted line), $\pi/6$ (dotted line), and 0 (dashed line).

Figure 2.60 displays the calculated profiles of the Lyman-7 line for the scaled laser amplitude $\varepsilon = 2$ for four different observation angles: $\pi/2$ (solid line), $\pi/3$ (dash-dotted line), $\pi/6$ (dotted line), and 0 (dashed line).

From figures 2.55–2.60 one can see the following. First, the Lyman-7 profiles are even more sensitive to the laser field than the Lyman-beta and Lyman-delta profiles. They are also sensitive to the direction of the observation. As the angle of the observation decreases from $\pi/2$ to 0, the half width at half maximum diminishes.

Second, there is again a clear distinction from the corresponding Lyman-7 profiles for the case of the one-mode laser field. For the latter case, as the scaled laser amplitude ε increases, the primary maximum of the satellites intensity moves away from the unperturbed position of the spectral line, and the greater the value of ε, the further away the primary maximum moves. However, in the case of the two-mode laser field, the primary maximum of the satellites intensities remains at the unperturbed position of the spectral line.

Third, again the peaks of the satellites intensities have a monotonically decreasing envelope. This differs from the case of the one-mode laser field.

Now we proceed to studying the angular dependence of the Lyman-9 profiles. Figure 2.61 shows its profiles for the scaled laser amplitude $\varepsilon = 0.031\,25$ for four different observation angles: $\pi/2$ (solid line), $\pi/3$ (dash-dotted line), $\pi/6$ (dotted line), and 0 (dashed line).

Figure 2.62 demonstrates the calculated profiles of the Lyman-9 line for the scaled laser amplitude $\varepsilon = 0.0625$ for four different observation angles: $\pi/2$ (solid line), $\pi/3$ (dash-dotted line), $\pi/6$ (dotted line), and 0 (dashed line).

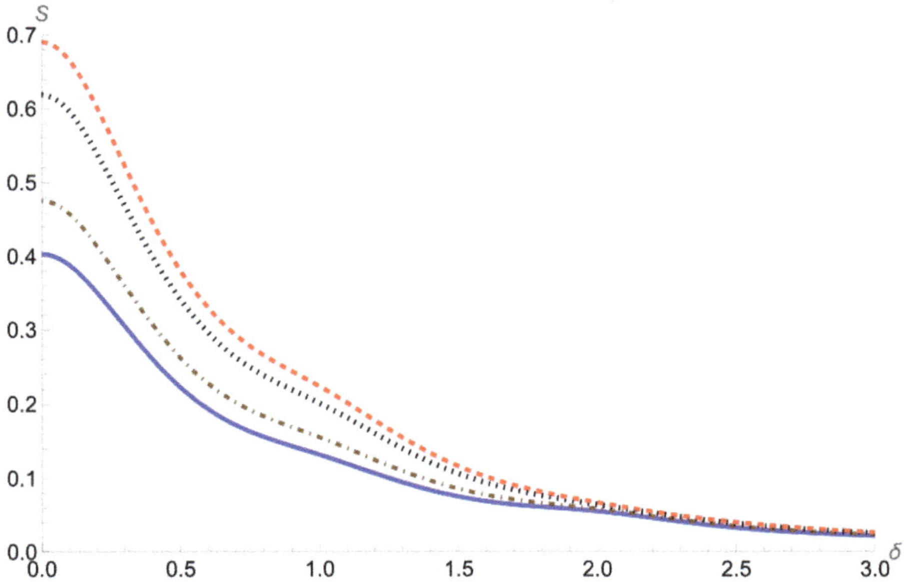

Figure 2.61. Calculated profiles of the Lyman-9 line for the scaled laser amplitude $\varepsilon = 0.031\ 25$ for four different observation angles: $\pi/2$ (solid line), $\pi/3$ (dash-dotted line), $\pi/6$ (dotted line), and 0 (dashed line).

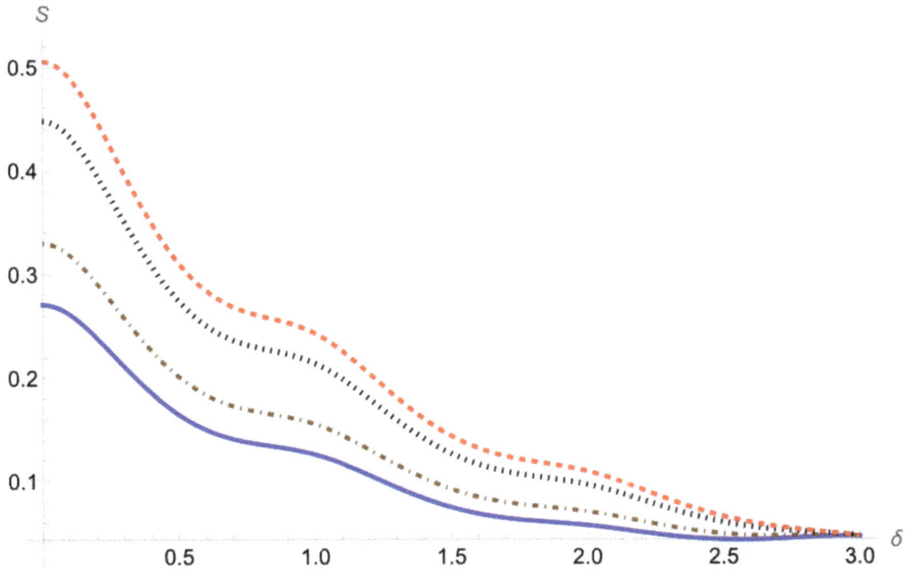

Figure 2.62. Calculated profiles of the Lyman-9 line for the scaled laser amplitude $\varepsilon = 0.0625$ for four different observation angles: $\pi/2$ (solid line), $\pi/3$ (dash-dotted line), $\pi/6$ (dotted line), and 0 (dashed line).

Figure 2.63 displays the calculated profiles of the Lyman-9 line for the scaled laser amplitude $\varepsilon = 0.125$ for four different observation angles: $\pi/2$ (solid line), $\pi/3$ (dash-dotted line), $\pi/6$ (dotted line), and 0 (dashed line).

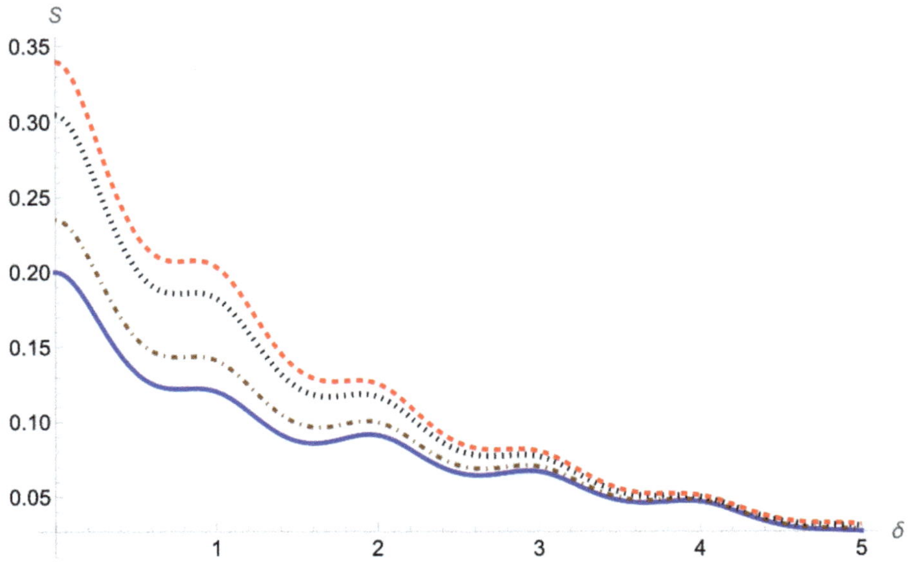

Figure 2.63. Calculated profiles of the Lyman-9 line for the scaled laser amplitude $\varepsilon = 0.125$ for four different observation angles: $\pi/2$ (solid line), $\pi/3$ (dash-dotted line), $\pi/6$ (dotted line), and 0 (dashed line).

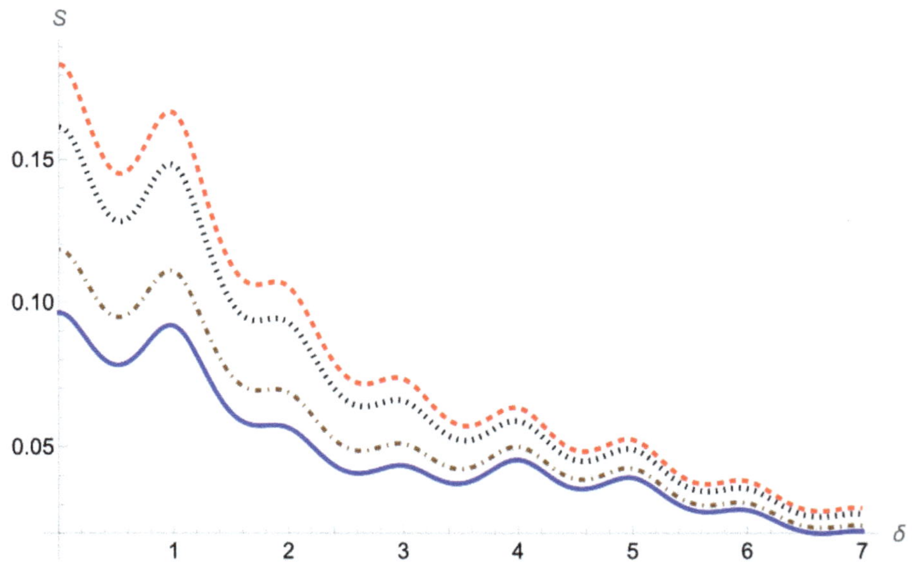

Figure 2.64. Calculated profiles of the Lyman-9 line for the scaled laser amplitude $\varepsilon = 0.25$ for four different observation angles: $\pi/2$ (solid line), $\pi/3$ (dash-dotted line), $\pi/6$ (dotted line), and 0 (dashed line).

Figure 2.64 shows the calculated profiles of the Lyman-9 line for the scaled laser amplitude $\varepsilon = 0.25$ for four different observation angles: $\pi/2$ (solid line), $\pi/3$ (dash-dotted line), $\pi/6$ (dotted line), and 0 (dashed line).

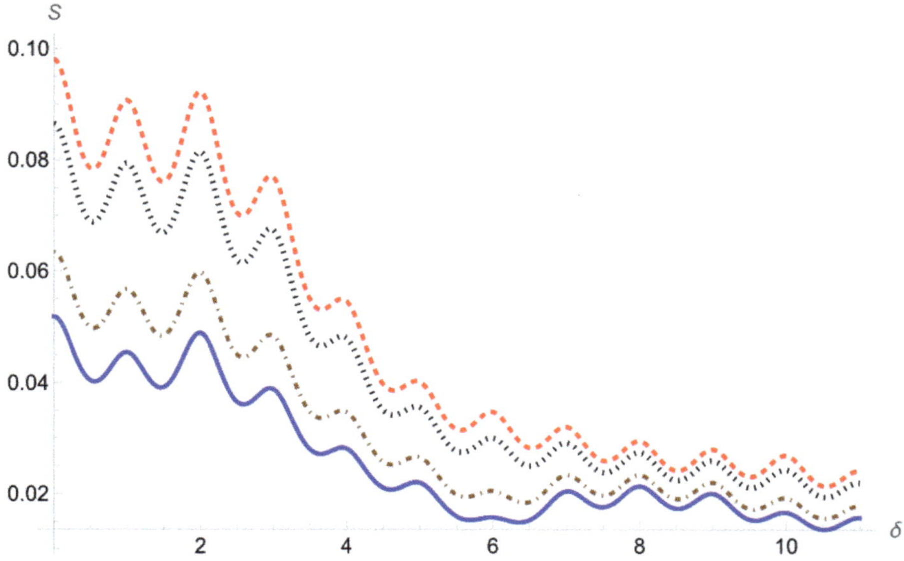

Figure 2.65. Calculated profiles of the Lyman-9 line for the scaled laser amplitude $\varepsilon = 0.5$ for four different observation angles: $\pi/2$ (solid line), $\pi/3$ (dash-dotted line), $\pi/6$ (dotted line), and 0 (dashed line).

Figure 2.65 demonstrates the calculated profiles of the Lyman-9 line for the scaled laser amplitude $\varepsilon = 0.5$ for four different observation angles: $\pi/2$ (solid line), $\pi/3$ (dash-dotted line), $\pi/6$ (dotted line), and 0 (dashed line).

Figure 2.66 displays the calculated profiles of the Lyman-9 line for the scaled laser amplitude $\varepsilon = 1$ for four different observation angles: $\pi/2$ (solid line), $\pi/3$ (dash-dotted line), $\pi/6$ (dotted line), and 0 (dashed line).

From figures 2.61–2.66 one can see the following. First of all, *the line Lyman-9 exhibits a qualitatively new feature* compared to the lines Lyman-7, Lyman-delta, and Lyman-beta. Namely, for the values of the scaled amplitude of the laser field $\varepsilon = 0.5$ and $\varepsilon = 1$, *the envelope of the satellites peaks is not monotonic anymore*. It starts by monotonically decreasing, then it reaches a minimum, and then exhibits a secondary maximum.

Second, the Lyman-9 profiles are significantly more sensitive to the laser field than the Lyman-beta, Lyman-delta, and Lyman-7 profiles. They are also sensitive to the direction of the observation. As the angle of the observation decreases from $\pi/2$ to 0, the half width at half maximum diminishes.

Now we proceed to studying the angular dependence of the Lyman-11 profiles. Figure 2.67 shows its profiles for the scaled laser amplitude $\varepsilon = 0.031\,25$ for four different observation angles: $\pi/2$ (solid line), $\pi/3$ (dash-dotted line), $\pi/6$ (dotted line), and 0 (dashed line).

Figure 2.68 demonstrates the calculated profiles of the Lyman-11 line for the scaled laser amplitude $\varepsilon = 0.0625$ for four different observation angles: $\pi/2$ (solid line), $\pi/3$ (dash-dotted line), $\pi/6$ (dotted line), and 0 (dashed line).

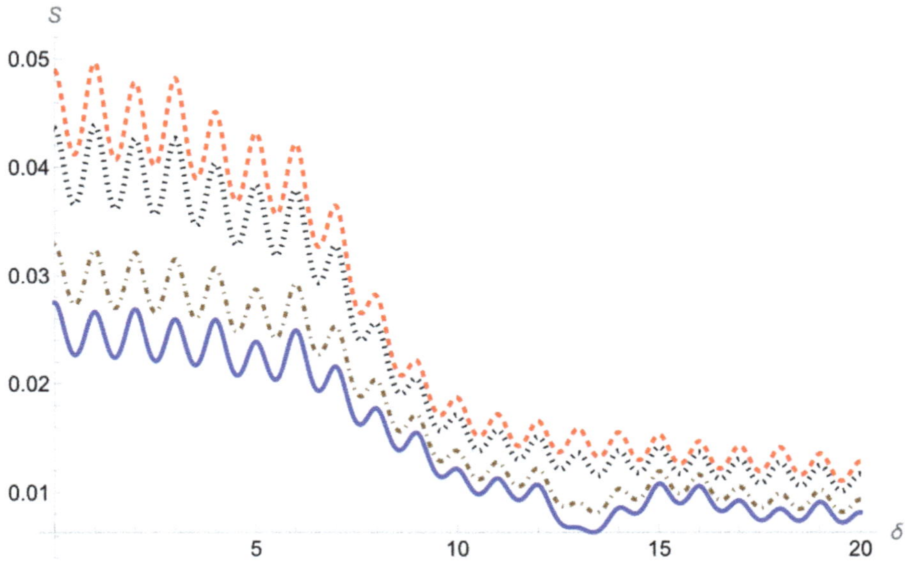

Figure 2.66. Calculated profiles of the Lyman-9 line for the scaled laser amplitude $\varepsilon = 1$ for four different observation angles: $\pi/2$ (solid line), $\pi/3$ (dash-dotted line), $\pi/6$ (dotted line), and 0 (dashed line).

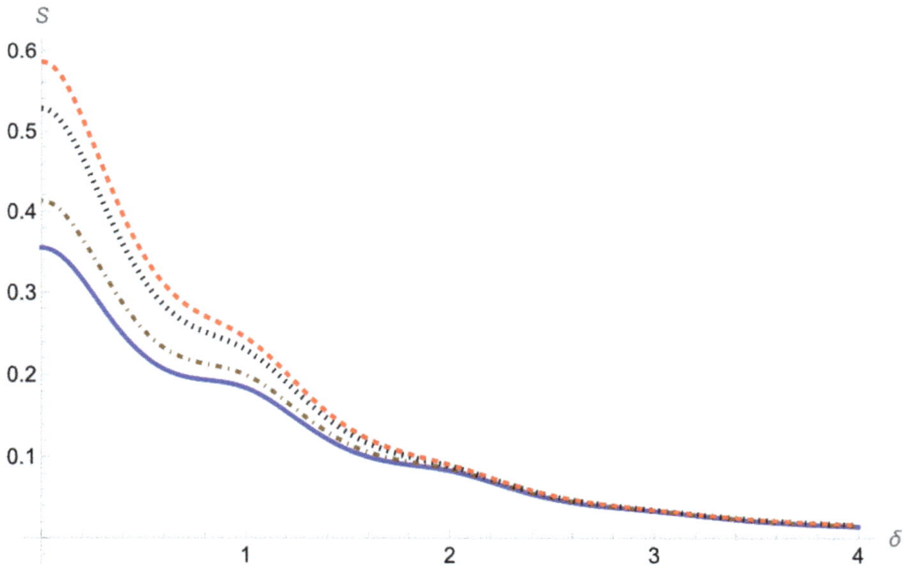

Figure 2.67. Calculated profiles of the Lyman-11 line for the scaled laser amplitude $\varepsilon = 0.031\ 25$ for four different observation angles: $\pi/2$ (solid line), $\pi/3$ (dash-dotted line), $\pi/6$ (dotted line), and 0 (dashed line).

Figure 2.69 displays the calculated profiles of the Lyman-11 line for the scaled laser amplitude $\varepsilon = 0.125$ for four different observation angles: $\pi/2$ (solid line), $\pi/3$ (dash-dotted line), $\pi/6$ (dotted line), and 0 (dashed line).

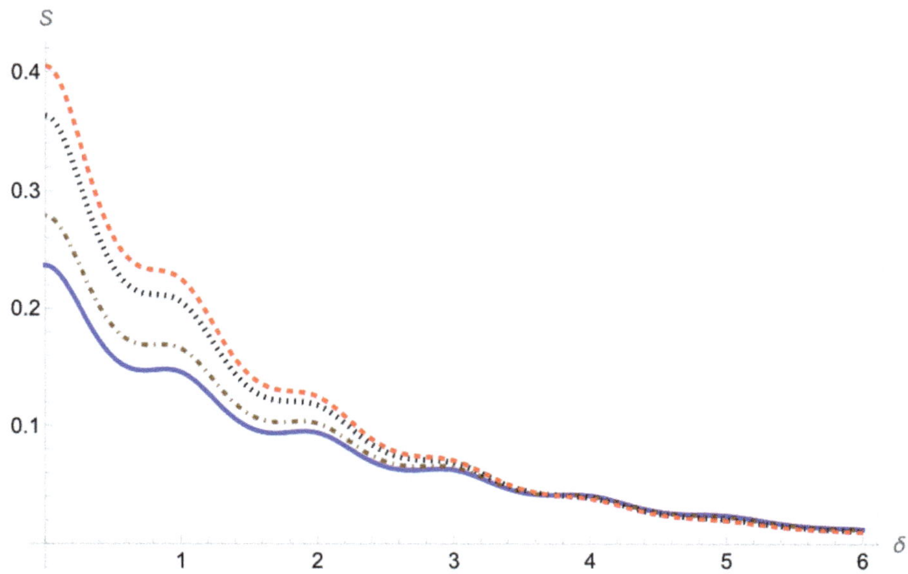

Figure 2.68. Calculated profiles of the Lyman-11 line for the scaled laser amplitude $\varepsilon = 0.0625$ for four different observation angles: $\pi/2$ (solid line), $\pi/3$ (dash-dotted line), $\pi/6$ (dotted line), and 0 (dashed line).

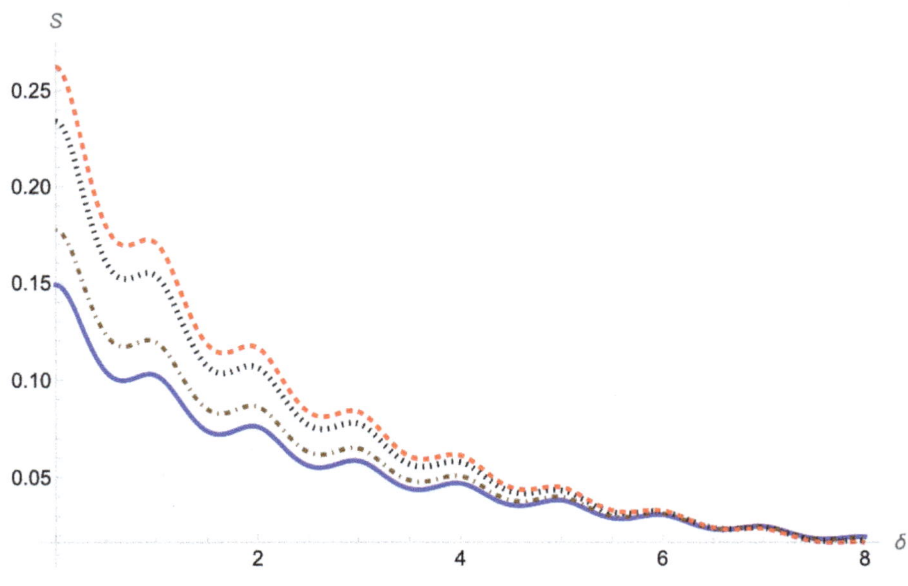

Figure 2.69. Calculated profiles of the Lyman-11 line for the scaled laser amplitude $\varepsilon = 0.125$ for four different observation angles: $\pi/2$ (solid line), $\pi/3$ (dash-dotted line), $\pi/6$ (dotted line), and 0 (dashed line).

Figure 2.70 shows the calculated profiles of the Lyman-11 line for the scaled laser amplitude $\varepsilon = 0.25$ for four different observation angles: $\pi/2$ (solid line), $\pi/3$ (dash-dotted line), $\pi/6$ (dotted line), and 0 (dashed line).

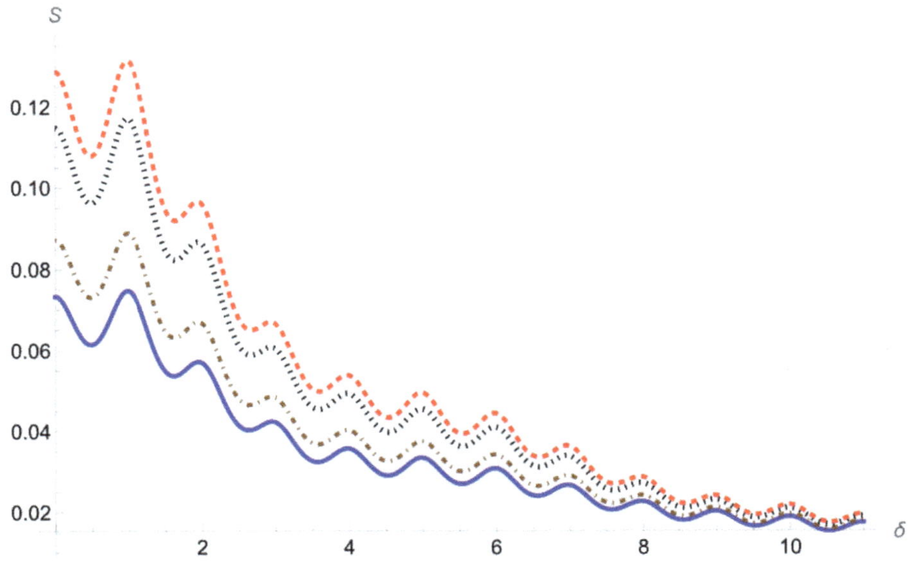

Figure 2.70. Calculated profiles of the Lyman-11 line for the scaled laser amplitude $\varepsilon = 0.25$ for four different observation angles: $\pi/2$ (solid line), $\pi/3$ (dash-dotted line), $\pi/6$ (dotted line), and 0 (dashed line).

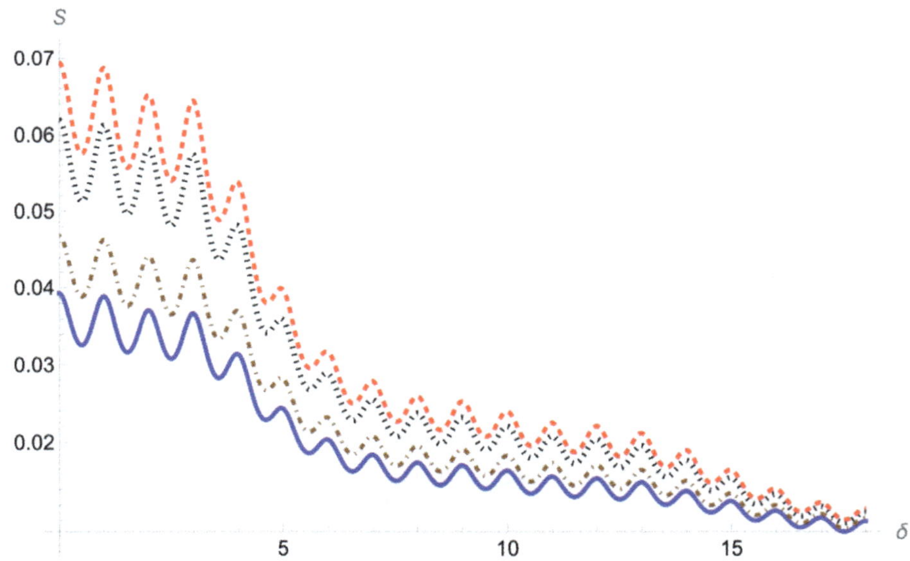

Figure 2.71. Calculated profiles of the Lyman-11 line for the scaled laser amplitude $\varepsilon = 0.5$ for four different observation angles: $\pi/2$ (solid line), $\pi/3$ (dash-dotted line), $\pi/6$ (dotted line), and 0 (dashed line).

Figure 2.71 demonstrates the calculated profiles of the Lyman-11 line for the scaled laser amplitude $\varepsilon = 0.5$ for four different observation angles: $\pi/2$ (solid line), $\pi/3$ (dash-dotted line), $\pi/6$ (dotted line), and 0 (dashed line).

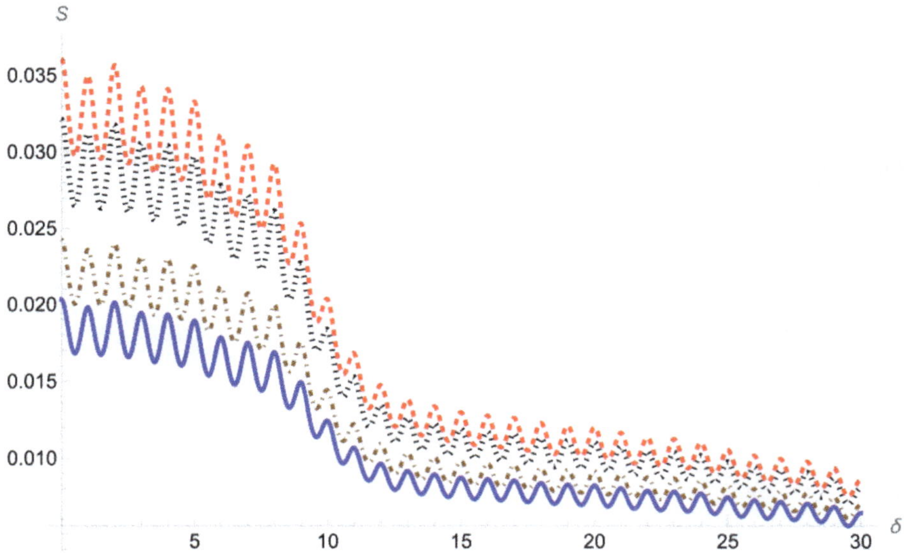

Figure 2.72. Calculated profiles of the Lyman-11 line for the scaled laser amplitude $\varepsilon = 0.5$ for four different observation angles: $\pi/2$ (solid line), $\pi/3$ (dash-dotted line), $\pi/6$ (dotted line), and 0 (dashed line).

Figure 2.72 shows the calculated profiles of the Lyman-11 line for the scaled laser amplitude $\varepsilon = 1$ for four different observation angles: $\pi/2$ (solid line), $\pi/3$ (dash-dotted line), $\pi/6$ (dotted line), and 0 (dashed line).

From figures 2.67–2.72 one can see the following. First, the remarkable feature exhibited by the line Lyman-9 does—the feature that the envelope of the satellites peaks is not monotonic—is not reproduced in the profiles of the line Lyman-11: the envelope of satellites shows the monotonic decrease.

Second, the Lyman-11 profiles are much more sensitive to the laser field than the Lyman-beta, Lyman-delta, Lyman-7, and Lyman-9 profiles. They are also sensitive to the direction of the observation. As the angle of the observation decreases from $\pi/2$ to 0, the half width at half maximum decreases.

Now we proceed to studying the angular dependence of the Balmer-beta profiles. Figure 2.73 shows its profiles for the scaled laser amplitude $\varepsilon = 0.125$ for four different observation angles: $\pi/2$ (solid line), $\pi/3$ (dash-dotted line), $\pi/6$ (dotted line), and 0 (dashed line).

Figure 2.74 demonstrates the calculated profiles of the Balmer-beta line for the scaled laser amplitude $\varepsilon = 25$ for four different observation angles: $\pi/2$ (solid line), $\pi/3$ (dash-dotted line), $\pi/6$ (dotted line), and 0 (dashed line).

Figure 2.75 displays the calculated profiles of the Balmer-beta line for the scaled laser amplitude $\varepsilon = 0.5$ for four different observation angles: $\pi/2$ (solid line), $\pi/3$ (dash-dotted line), $\pi/6$ (dotted line), and 0 (dashed line).

Figure 2.76 shows the calculated profiles of the Balmer-beta line for the scaled laser amplitude $\varepsilon = 1$ for four different observation angles: $\pi/2$ (solid line), $\pi/3$ (dash-dotted line), $\pi/6$ (dotted line), and 0 (dashed line).

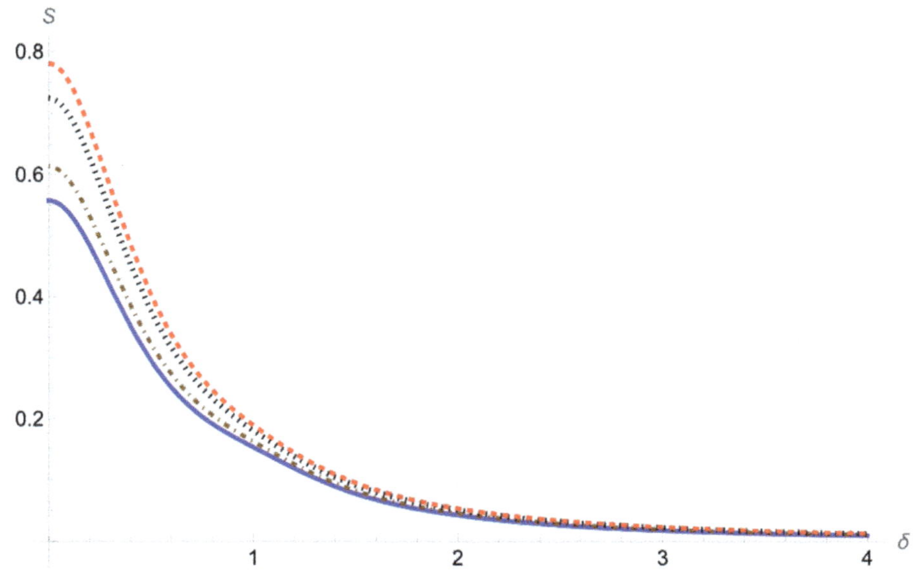

Figure 2.73. Calculated profiles of the Balmer-beta line for the scaled laser amplitude $\varepsilon = 0.125$ for four different observation angles: $\pi/2$ (solid line), $\pi/3$ (dash-dotted line), $\pi/6$ (dotted line), and 0 (dashed line).

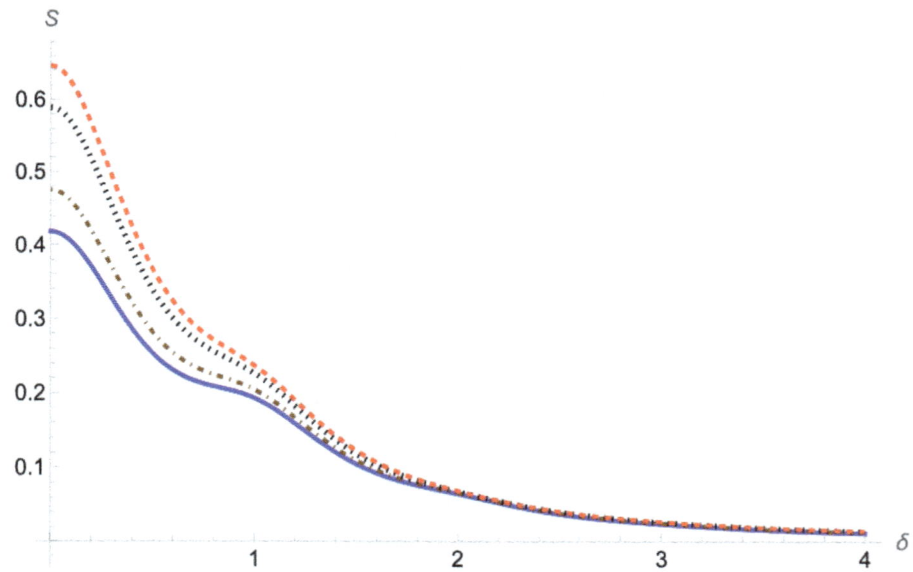

Figure 2.74. Calculated profiles of the Balmer-beta line for the scaled laser amplitude $\varepsilon = 0.25$ for four different observation angles: $\pi/2$ (solid line), $\pi/3$ (dash-dotted line), $\pi/6$ (dotted line), and 0 (dashed line).

Figure 2.77 demonstrates the calculated profiles of the Balmer-beta line for the scaled laser amplitude $\varepsilon = 2$ for four different observation angles: $\pi/2$ (solid line), $\pi/3$ (dash-dotted line), $\pi/6$ (dotted line), and 0 (dashed line).

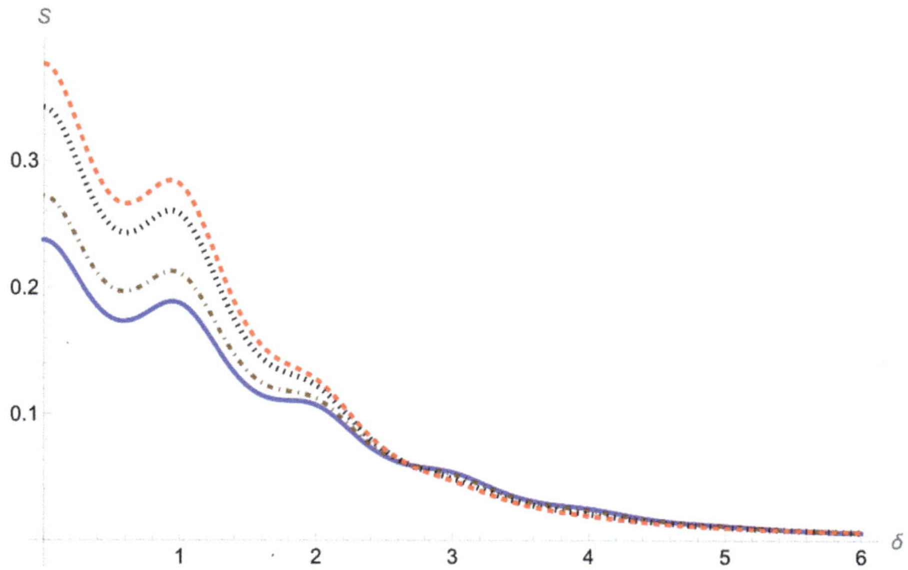

Figure 2.75. Calculated profiles of the Balmer-beta line for the scaled laser amplitude $\varepsilon = 0.5$ for four different observation angles: $\pi/2$ (solid line), $\pi/3$ (dash-dotted line), $\pi/6$ (dotted line), and 0 (dashed line).

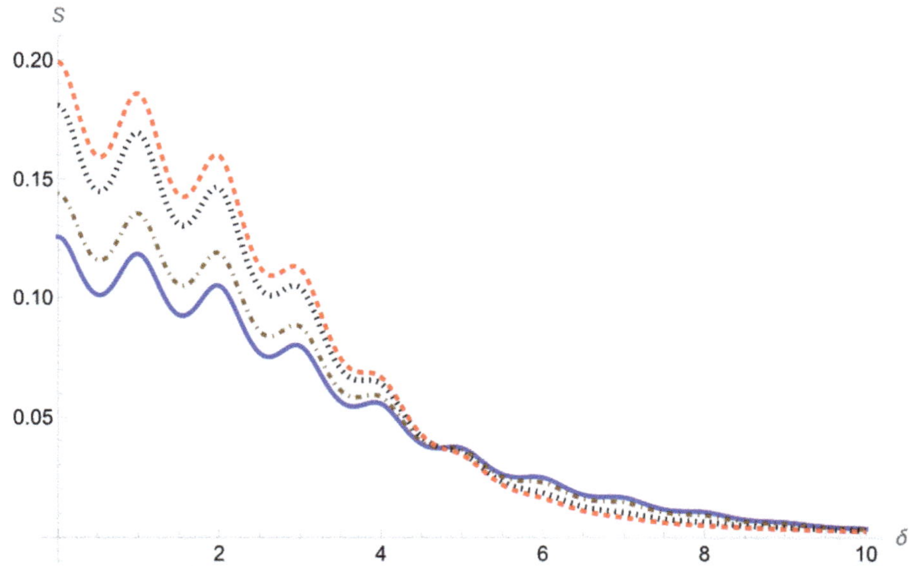

Figure 2.76. Calculated profiles of the Balmer-beta line for the scaled laser amplitude $\varepsilon = 1$ for four different observation angles: $\pi/2$ (solid line), $\pi/3$ (dash-dotted line), $\pi/6$ (dotted line), and 0 (dashed line).

Figure 2.78 displays the calculated profiles of the Balmer-beta line for the scaled laser amplitude $\varepsilon = 4$ for four different observation angles: $\pi/2$ (solid line), $\pi/3$ (dash-dotted line), $\pi/6$ (dotted line), and 0 (dashed line).

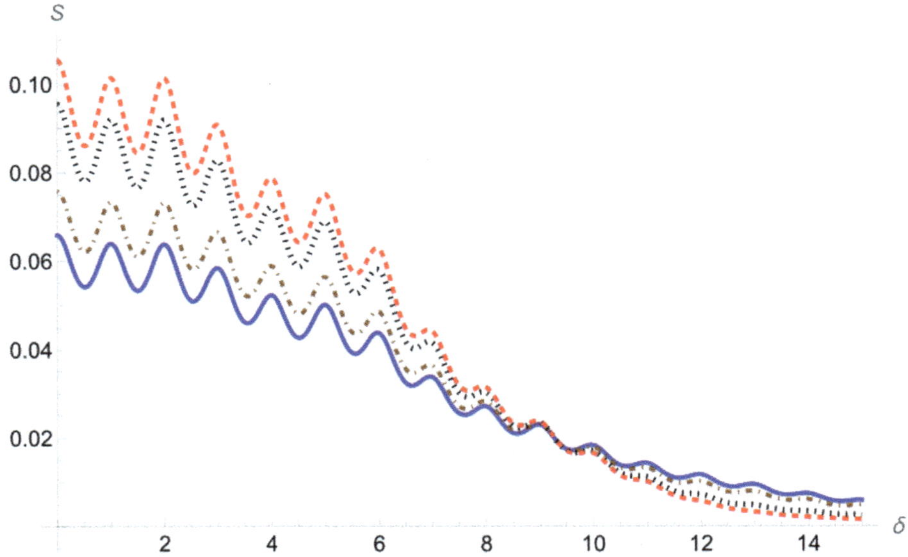

Figure 2.77. Calculated profiles of the Balmer-beta line for the scaled laser amplitude $\varepsilon = 2$ for four different observation angles: $\pi/2$ (solid line), $\pi/3$ (dash-dotted line), $\pi/6$ (dotted line), and 0 (dashed line).

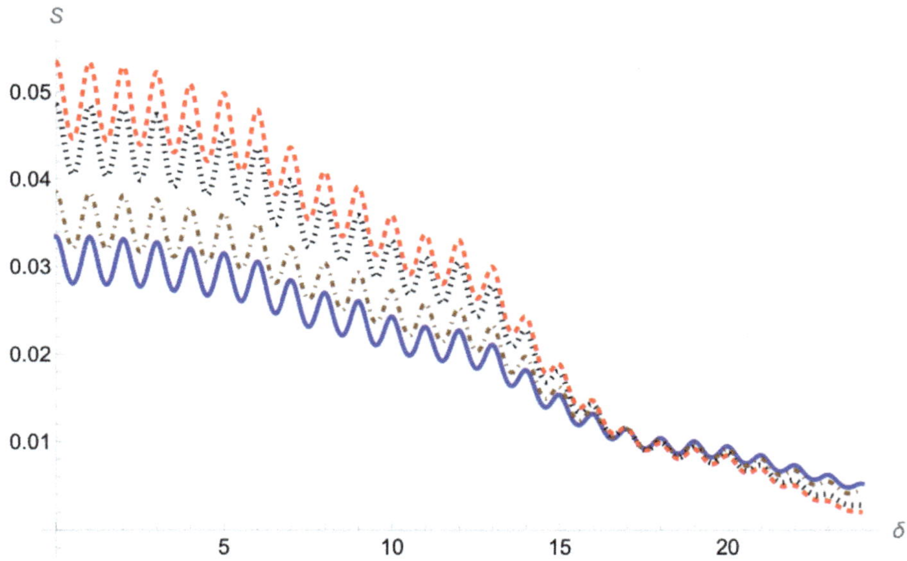

Figure 2.78. Calculated profiles of the Balmer-beta line for the scaled laser amplitude $\varepsilon = 4$ for four different observation angles: $\pi/2$ (solid line), $\pi/3$ (dash-dotted line), $\pi/6$ (dotted line), and 0 (dashed line).

From figures 2.73–2.78 one can see the following. First, the Balmer-beta profiles relatively sensitive to the amplitude laser field. They are also sensitive to the direction of the observation. As the angle of the observation decreases from $\pi/2$ to 0, the half width at half maximum diminishes.

Second, there is a clear distinction from the corresponding Balmer-beta profiles for the case of the one-mode laser field. For the latter case, as the scaled laser amplitude ε increases, the primary maximum of the satellites intensity moves away from the unperturbed position of the spectral line, and the greater the value of ε, the further away the primary maximum moves. However, in the case of the two-mode laser field, the primary maximum of the satellites intensities remains at the unperturbed position of the spectral line.

Third, the peaks of the satellites intensities have a monotonically decreasing envelope. This differs from the case of the one-mode laser field.

Finally, we proceed to studying the angular dependence of the Balmer-delta profiles. Figure 2.79 shows its profiles for the scaled laser amplitude $\varepsilon = 0.0625$ for four different observation angles: $\pi/2$ (solid line), $\pi/3$ (dash-dotted line), $\pi/6$ (dotted line), and 0 (dashed line).

Figure 2.80 demonstrates the calculated profiles of the Balmer-delta line for the scaled laser amplitude $\varepsilon = 0.125$ for four different observation angles: $\pi/2$ (solid line), $\pi/3$ (dash-dotted line), $\pi/6$ (dotted line), and 0 (dashed line).

Figure 2.81 displays the calculated profiles of the Balmer-delta line for the scaled laser amplitude $\varepsilon = 0.25$ for four different observation angles: $\pi/2$ (solid line), $\pi/3$ (dash-dotted line), $\pi/6$ (dotted line), and 0 (dashed line).

Figure 2.82 shows the calculated profiles of the Balmer-delta line for the scaled laser amplitude $\varepsilon = 0.5$ for four different observation angles: $\pi/2$ (solid line), $\pi/3$ (dash-dotted line), $\pi/6$ (dotted line), and 0 (dashed line).

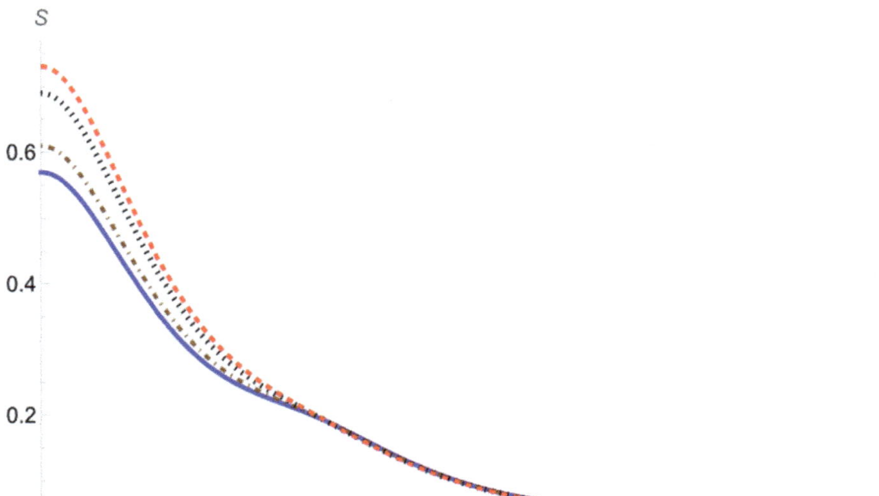

Figure 2.79. Calculated profiles of the Balmer-delta line for the scaled laser amplitude $\varepsilon = 0.0625$ for four different observation angles: $\pi/2$ (solid line), $\pi/3$ (dash-dotted line), $\pi/6$ (dotted line), and 0 (dashed line).

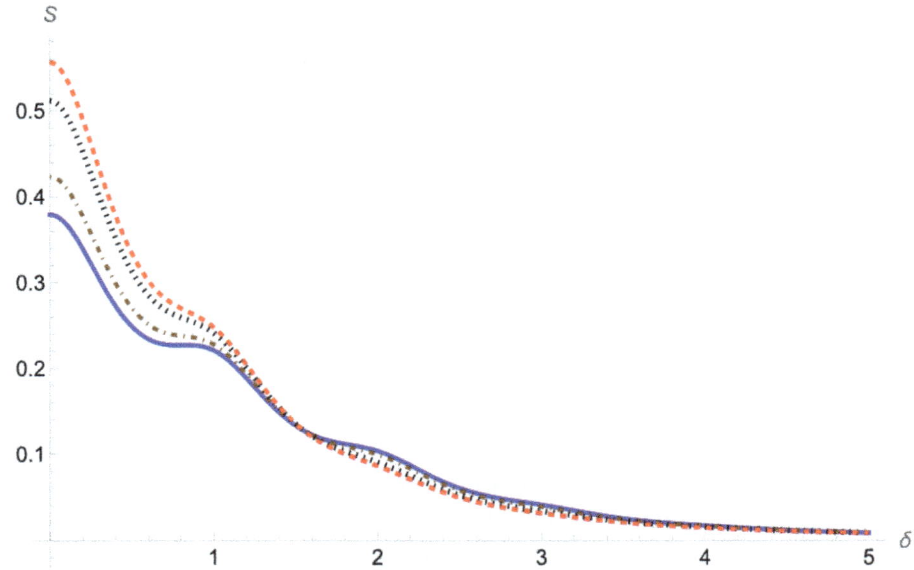

Figure 2.80. Calculated profiles of the Balmer-delta line for the scaled laser amplitude $\varepsilon = 0.125$ for four different observation angles: $\pi/2$ (solid line), $\pi/3$ (dash-dotted line), $\pi/6$ (dotted line), and 0 (dashed line).

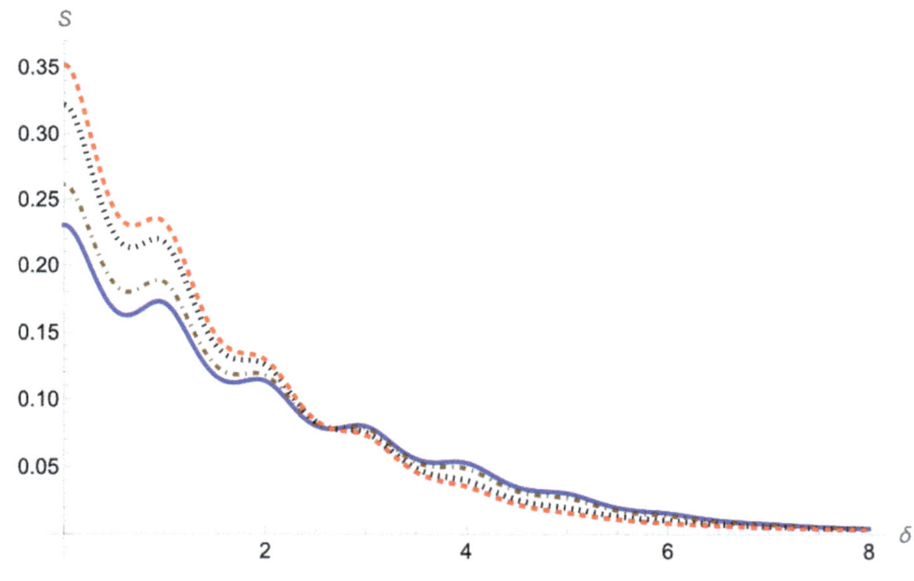

Figure 2.81. Calculated profiles of the Balmer-delta line for the scaled laser amplitude $\varepsilon = 0.25$ for four different observation angles: $\pi/2$ (solid line), $\pi/3$ (dash-dotted line), $\pi/6$ (dotted line), and 0 (dashed line).

Figure 2.83 demonstrates the calculated profiles of the Balmer-delta line for the scaled laser amplitude $\varepsilon = 1$ for four different observation angles: $\pi/2$ (solid line), $\pi/3$ (dash-dotted line), $\pi/6$ (dotted line), and 0 (dashed line).

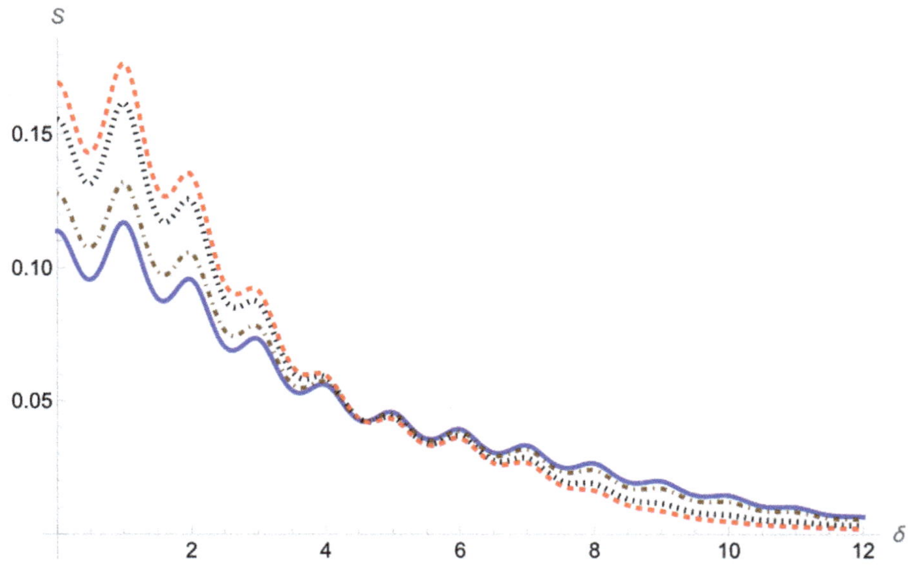

Figure 2.82. Calculated profiles of the Balmer-delta line for the scaled laser amplitude $\varepsilon = 0.5$ for four different observation angles: $\pi/2$ (solid line), $\pi/3$ (dash-dotted line), $\pi/6$ (dotted line), and 0 (dashed line).

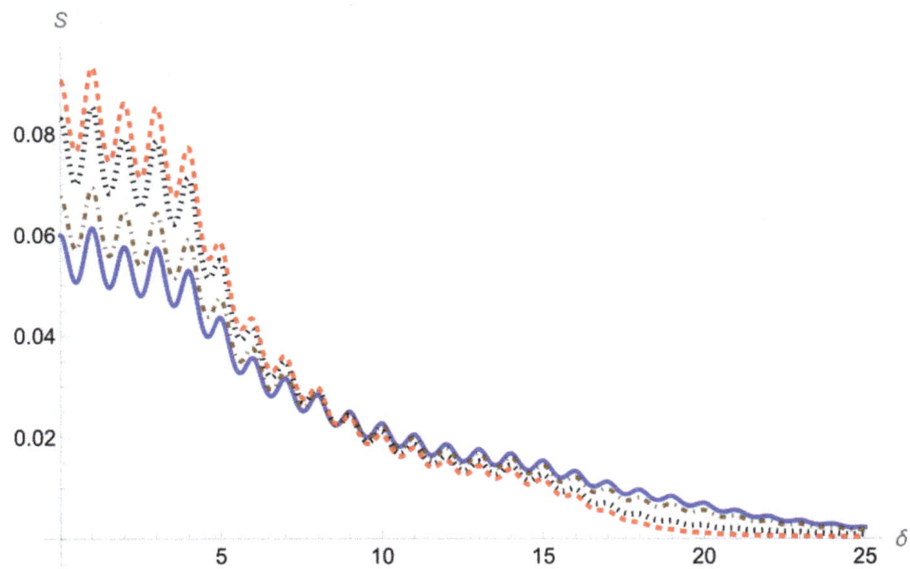

Figure 2.83. Calculated profiles of the Balmer-delta line for the scaled laser amplitude $\varepsilon = 1$ for four different observation angles: $\pi/2$ (solid line), $\pi/3$ (dash-dotted line), $\pi/6$ (dotted line), and 0 (dashed line).

Figure 2.84 displays the calculated profiles of the Balmer-delta line for the scaled laser amplitude $\varepsilon = 2$ for four different observation angles: $\pi/2$ (solid line), $\pi/3$ (dash-dotted line), $\pi/6$ (dotted line), and 0 (dashed line).

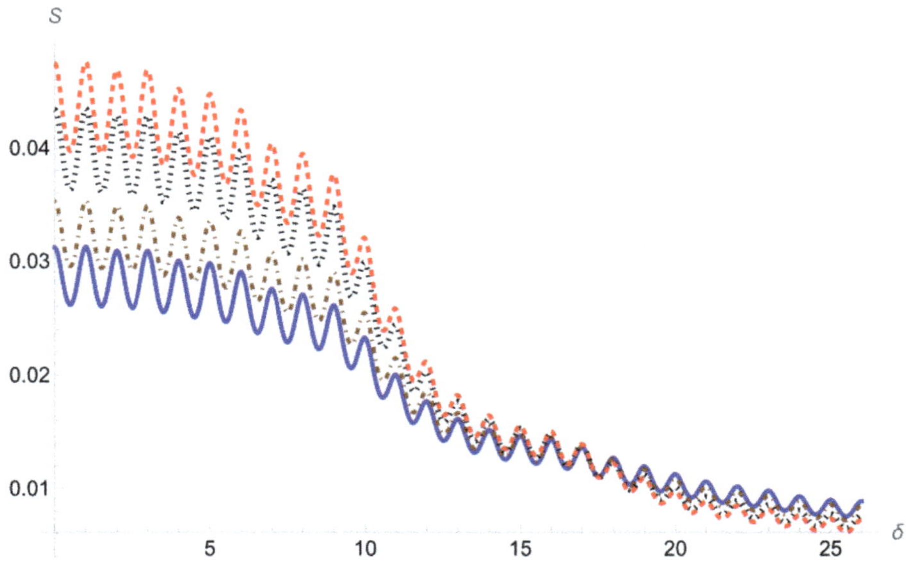

Figure 2.84. Calculated profiles of the Balmer-delta line for the scaled laser amplitude $\varepsilon = 1$ for four different observation angles: $\pi/2$ (solid line), $\pi/3$ (dash-dotted line), $\pi/6$ (dotted line), and 0 (dashed line).

From figures 2.79–2.84 one can see the following. First, the Balmer-delta profiles are significantly more sensitive to the laser field compared to the Balmer-beta profiles. They are also sensitive to the direction of the observation. As the angle of the observation decreases from $\pi/2$ to 0, the half width at half maximum diminishes.

Second, there is again a clear distinction from the corresponding Balmer-delta profiles for the case of the one-mode laser field. For the latter case, as the scaled laser amplitude ε increases, the primary maximum of the satellites intensity moves away from the unperturbed position of the spectral line, and the greater the value of ε, the further away the primary maximum moves. However, in the case of the two-mode laser field, the primary maximum of the satellites intensities remains at the unperturbed position of the spectral line.

Third, again the peaks of the satellites intensities have a monotonically decreasing envelope. This differs from the case of the one-mode laser field.

Summarizing the section 2.2, we could provide the following application of the results for diagnosing the two-mode laser field in plasmas, where due to nonlinear processes, the amplitude of the electromagnetic field at the laser frequency differs from the amplitude of the incoming laser field. The one-dimensional two-mode monochromatic electric field can create numerous satellites of hydrogenic spectral lines. By trying several hydrogenic lines, an experimentalist can find the line most sensitive to the relevant range of the laser field in the particular experiments. In distinction to the case of the one-mode laser field, where most of the attention should be given to the location and the intensity of the primary maximum and of the secondary maximum (since those maxima are away from the unperturbed location of the corresponding spectral line), in the case of the two-mode laser field, there is only one maximum and it is situated at the unperturbed location of the corresponding spectral line—except for the Lyman-9 line. Therefore, the best choice is the Lyman-9 line (if possible): the amplitude of the two-mode laser field can be easily determined from the location of the secondary maximum (that shows up at sufficiently large field). As for the other spectral lines, it would be necessary to measure profiles of at least two spectral lines: at the same laser amplitude, they have different halfwidths of the envelopes of the satellites peaks—this would allow the experimental determination of the amplitude of the two-mode laser field despite the presence of other line broadening mechanisms in plasmas. For enhancing the reliability of the determination of the amplitude of the electromagnetic field at the laser frequency inside the plasma, it is a good idea to compare the profiles of the same spectral line observed at different angles—thus taking the advantage of the directional effects described above.

2.3 Satellites under the one-dimensional multi-mode monochromatic electric field

A one-dimensional multi-mode monochromatic electric field can be written as follows

$$\mathbf{E}(t) = \sum_{j=1}^{N} \mathbf{E}_j \cos(\omega t + \varphi_j), \tag{2.10}$$

where the number of modes is $N \gg 1$. The corresponding profile of a Stark component of a hydrogen line was calculated analytically by Lifshitz [4]:

$$S_{L,\,\text{profile}}(\Delta\omega/\omega) = \sum_{p=-\infty}^{\infty} K(a, p)\,(\Delta\omega/\omega - p). \tag{2.11}$$

In equation (2.11),

$$a = (X\varepsilon)^2, \;\; \varepsilon = 3\hbar E_0/(2m_e e\omega), \;\; E_0 = \left(\sum_{j=1}^{N} E_j^2\right)^{1/2}, \tag{2.12}$$

where X was defined in equation (2.2). Further, in equation (2.11)

$$K(a, p) = I_{|p|}(a/2)\exp(-a/2), \tag{2.13}$$

where $I_{|p|}(a/2)$ are the modified Bessel functions.

The profile of a multicomponent hydrogenic spectral line can be represented in the form:

$$S(\Delta\omega/\omega) = \sum_{p=-\infty}^{+\infty} L(p, \varepsilon)\delta(\Delta\omega/\omega) - p), \tag{2.14}$$

where

$$L(p, \varepsilon) = \left[f_0(\theta)\delta_{p0} + 2\sum_{k=1}^{k_{\max}} f_k(\theta)K(a, p)\right]/(f_0 + 2\Sigma f_k). \tag{2.15}$$

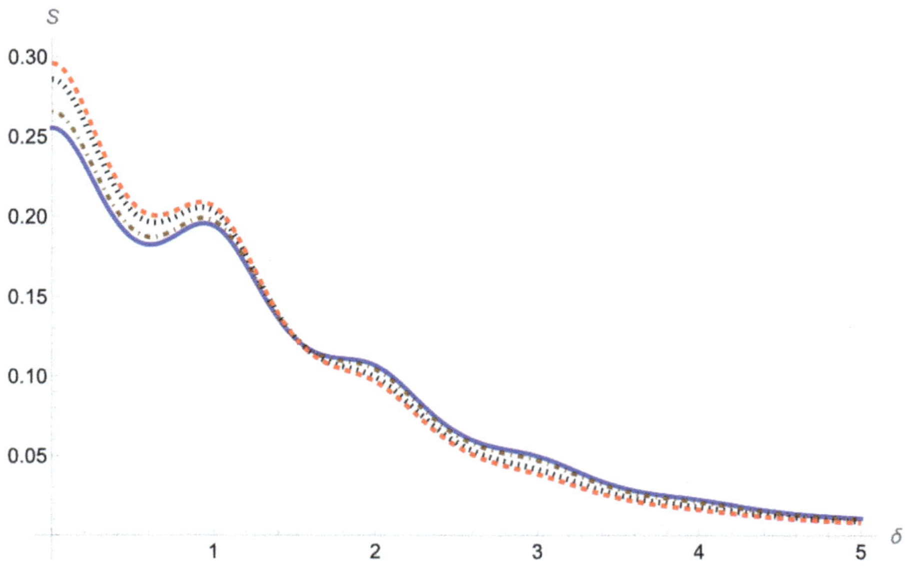

Figure 2.85. Calculated profiles of the Lyman-beta line for the scaled laser amplitude $\varepsilon = 0.5$ for four different observation angles: $\pi/2$ (solid line), $\pi/3$ (dash-dotted line), $\pi/6$ (dotted line), and 0 (dashed line).

Below we illustrate the angular dependence of profiles of the hydrogenic spectral lines, calculated by equations (2.14) and (2.15), that are the most useful for spectroscopic diagnostics of plasmas. For obtaining continuous profiles we assigned to each satellite the Lorentzian shape of the half width at half maximum equal to $\omega/4$, as in sections 2.1 and 2.2.

We start from the Lyman-beta line. Figure 2.85 shows its profile for the scaled laser amplitude $\varepsilon = 0.5$ for four different observation angles: $\pi/2$ (solid line), $\pi/3$ (dash-dotted line), $\pi/6$ (dotted line), and 0 (dashed line).

Figure 2.86 demonstrates the calculated profiles of the Lyman-beta line for the scaled laser amplitude $\varepsilon = 1$ for four different observation angles: $\pi/2$ (solid line), $\pi/3$ (dash-dotted line), $\pi/6$ (dotted line), and 0 (dashed line).

Figure 2.87 displays the calculated profiles of the Lyman-beta line for the scaled laser amplitude $\varepsilon = 2$ for four different observation angles: $\pi/2$ (solid line), $\pi/3$ (dash-dotted line), $\pi/6$ (dotted line), and 0 (dashed line).

Figure 2.88 shows the calculated profiles of the Lyman-beta line for the scaled laser amplitude $\varepsilon = 4$ for four different observation angles: $\pi/2$ (solid line), $\pi/3$ (dash-dotted line), $\pi/6$ (dotted line), and 0 (dashed line).

Figure 2.89 demonstrates the calculated profiles of the Lyman-beta line for the scaled laser amplitude $\varepsilon = 8$ for four different observation angles: $\pi/2$ (solid line), $\pi/3$ (dash-dotted line), $\pi/6$ (dotted line), and 0 (dashed line).

Figure 2.90 displays the calculated profiles of the Lyman-beta line for the scaled laser amplitude $\varepsilon = 16$ for four different observation angles: $\pi/2$ (solid line), $\pi/3$ (dash-dotted line), $\pi/6$ (dotted line), and 0 (dashed line).

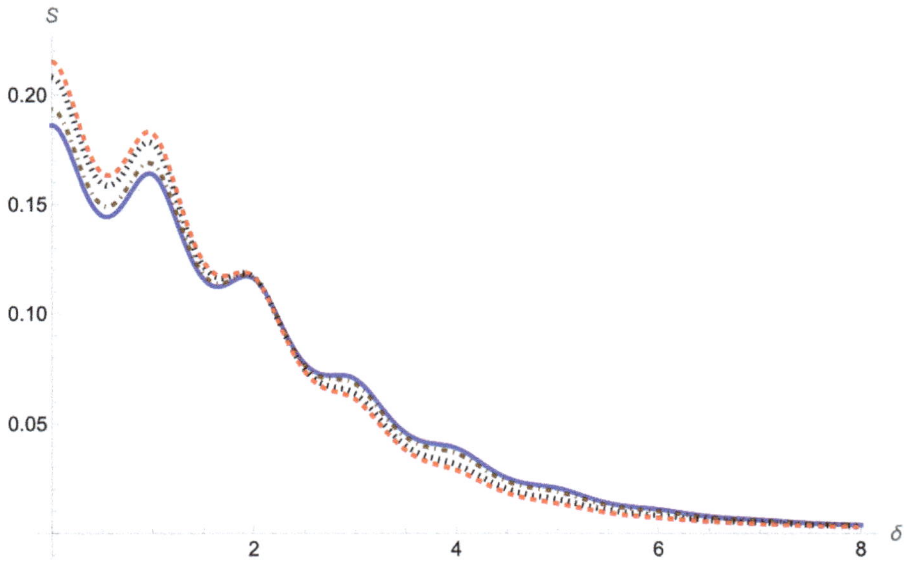

Figure 2.86. Calculated profiles of the Lyman-beta line for the scaled laser amplitude $\varepsilon = 1$ for four different observation angles: $\pi/2$ (solid line), $\pi/3$ (dash-dotted line), $\pi/6$ (dotted line), and 0 (dashed line).

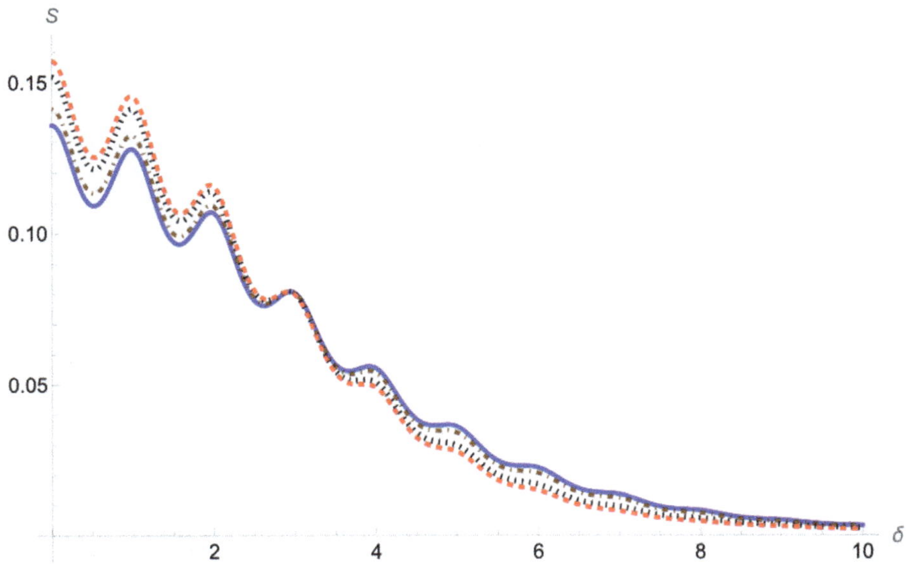

Figure 2.87. Calculated profiles of the Lyman-beta line for the scaled laser amplitude $\varepsilon = 2$ for four different observation angles: $\pi/2$ (solid line), $\pi/3$ (dash-dotted line), $\pi/6$ (dotted line), and 0 (dashed line).

From figures 2.85–2.90 one can see the following. First, the Lyman-beta profiles are sensitive to the direction of the observation. As the angle of the observation decreases from $\pi/2$ to 0, the half width at half maximum diminishes. However, this decrease is not so pronounced as for the corresponding profiles under the two-mode

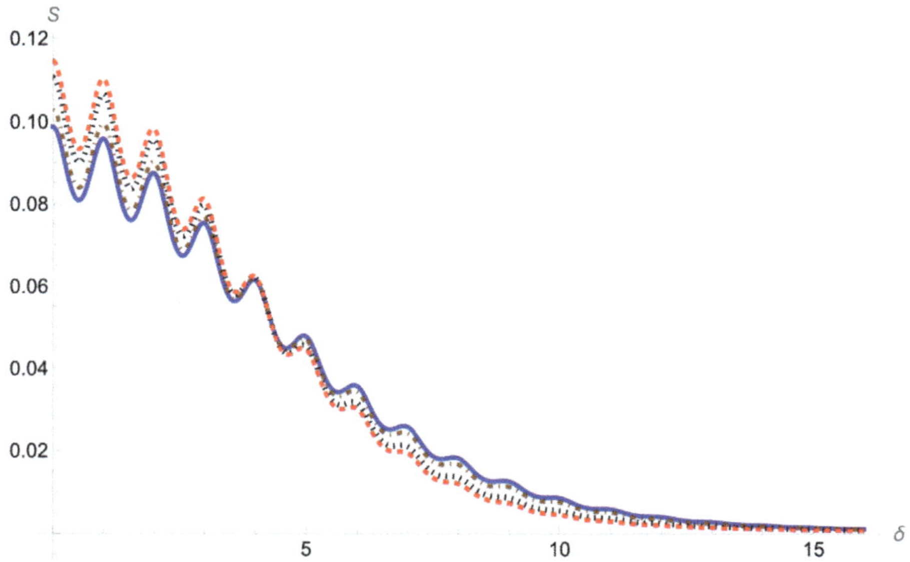

Figure 2.88. Calculated profiles of the Lyman-beta line for the scaled laser amplitude $\varepsilon = 4$ for four different observation angles: $\pi/2$ (solid line), $\pi/3$ (dash-dotted line), $\pi/6$ (dotted line), and 0 (dashed line).

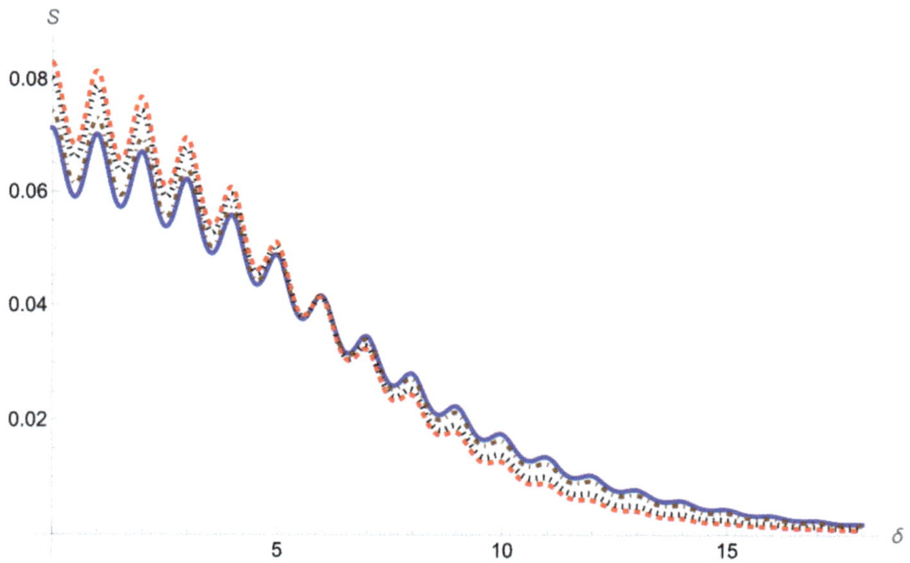

Figure 2.89. Calculated profiles of the Lyman-beta line for the scaled laser amplitude $\varepsilon = 8$ for four different observation angles: $\pi/2$ (solid line), $\pi/3$ (dash-dotted line), $\pi/6$ (dotted line), and 0 (dashed line).

laser field from section 2.2. *This is an important feature distinguishing the case of the multi-mode laser field from the case of the two-mode laser field.*

Second, the peaks of the satellites intensities have a monotonically decreasing envelope. This differs from the case of the one-mode laser field.

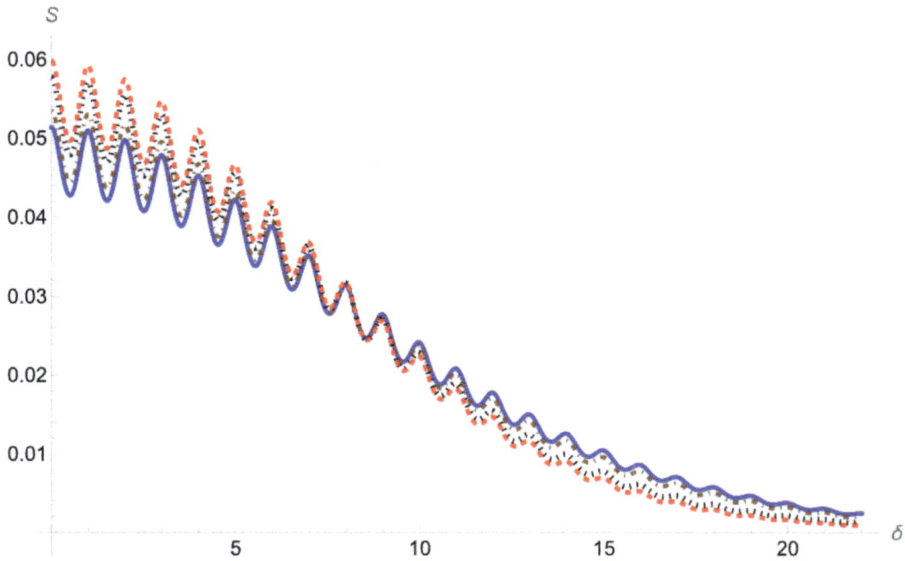

Figure 2.90. Calculated profiles of the Lyman-beta line for the scaled laser amplitude $\varepsilon = 16$ for four different observation angles: $\pi/2$ (solid line), $\pi/3$ (dash-dotted line), $\pi/6$ (dotted line), and 0 (dashed line).

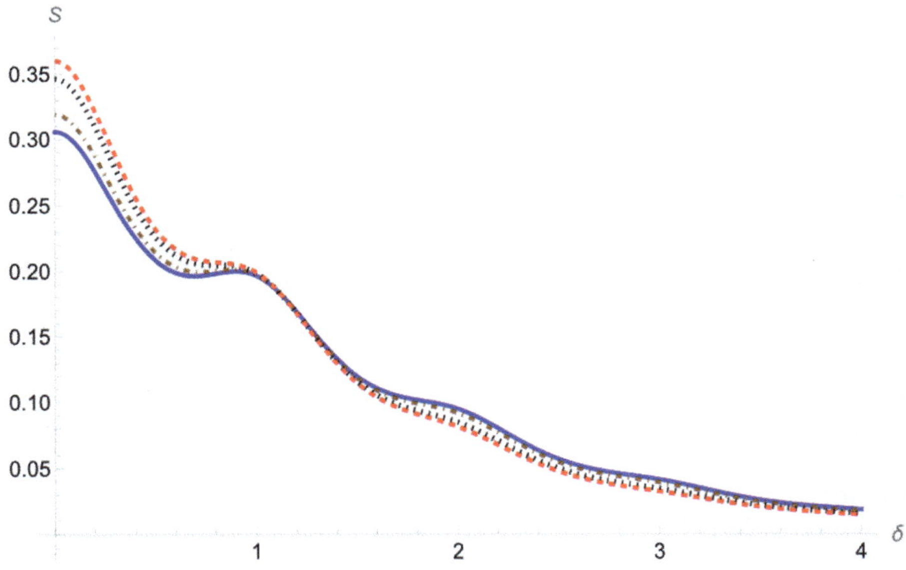

Figure 2.91. Calculated profiles of the Lyman-delta line for the scaled laser amplitude $\varepsilon = 0.125$ for four different observation angles: $\pi/2$ (solid line), $\pi/3$ (dash-dotted line), $\pi/6$ (dotted line), and 0 (dashed line).

Now we proceed to studying the angular dependence of the Lyman-delta profiles. Figure 2.91 shows its profiles for the scaled laser amplitude $\varepsilon = 0.125$ for four different observation angles: $\pi/2$ (solid line), $\pi/3$ (dash-dotted line), $\pi/6$ (dotted line), and 0 (dashed line).

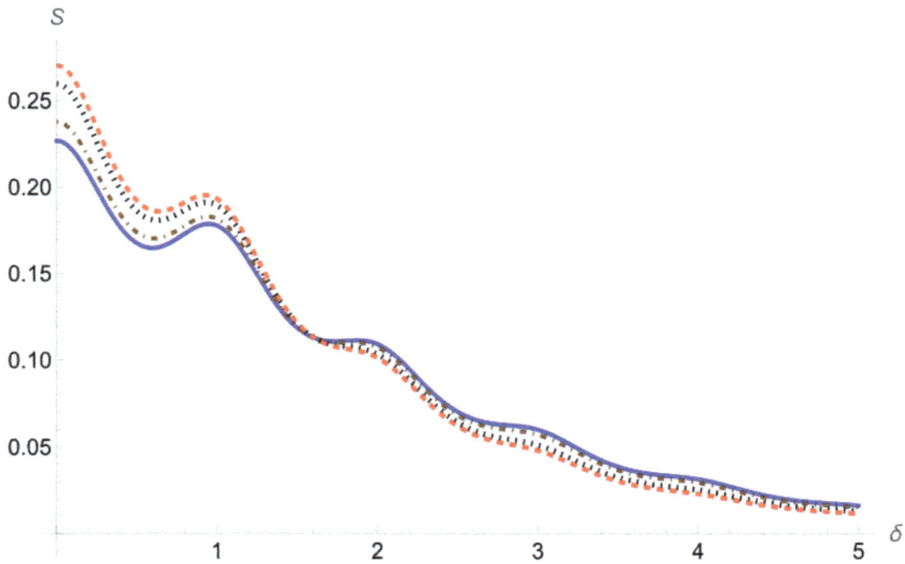

Figure 2.92. Calculated profiles of the Lyman-delta line for the scaled laser amplitude $\varepsilon = 0.25$ for four different observation angles: $\pi/2$ (solid line), $\pi/3$ (dash-dotted line), $\pi/6$ (dotted line), and 0 (dashed line).

Figure 2.92 demonstrates the calculated profiles of the Lyman-delta line for the scaled laser amplitude $\varepsilon = 0.25$ for four different observation angles: $\pi/2$ (solid line), $\pi/3$ (dash-dotted line), $\pi/6$ (dotted line), and 0 (dashed line).

Figure 2.93 displays the calculated profiles of the Lyman-delta line for the scaled laser amplitude $\varepsilon = 0.5$ for four different observation angles: $\pi/2$ (solid line), $\pi/3$ (dash-dotted line), $\pi/6$ (dotted line), and 0 (dashed line).

Figure 2.94 shows the calculated profiles of the Lyman-delta line for the scaled laser amplitude $\varepsilon = 1$ for four different observation angles: $\pi/2$ (solid line), $\pi/3$ (dash-dotted line), $\pi/6$ (dotted line), and 0 (dashed line).

Figure 2.95 demonstrates the calculated profiles of the Lyman-delta line for the scaled laser amplitude $\varepsilon = 2$ for four different observation angles: $\pi/2$ (solid line), $\pi/3$ (dash-dotted line), $\pi/6$ (dotted line), and 0 (dashed line).

Figure 2.96 displays the calculated profiles of the Lyman-delta line for the scaled laser amplitude $\varepsilon = 4$ for four different observation angles: $\pi/2$ (solid line), $\pi/3$ (dash-dotted line), $\pi/6$ (dotted line), and 0 (dashed line).

From figures 2.91–2.96 one can see the following. First, the Lyman-delta profiles are slightly more sensitive to the laser field compared to the corresponding Lyman-beta profiles.

Second, the Lyman-delta profiles are sensitive to the direction of the observation. As the angle of the observation decreases from $\pi/2$ to 0, the half width at half maximum diminishes. However, as for the corresponding profiles of the Lyman-beta line, this decrease is not so pronounced as for the corresponding profiles under the two-mode laser field from section 2.2. *This is an important feature distinguishing the case of the multi-mode laser field from the case of the two-mode laser field.*

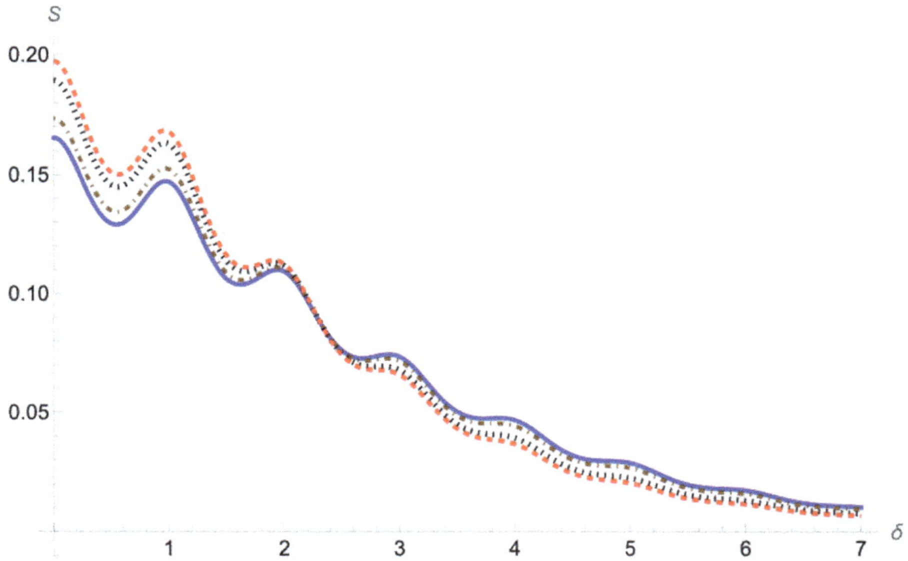

Figure 2.93. Calculated profiles of the Lyman-delta line for the scaled laser amplitude $\varepsilon = 0.5$ for four different observation angles: $\pi/2$ (solid line), $\pi/3$ (dash-dotted line), $\pi/6$ (dotted line), and 0 (dashed line).

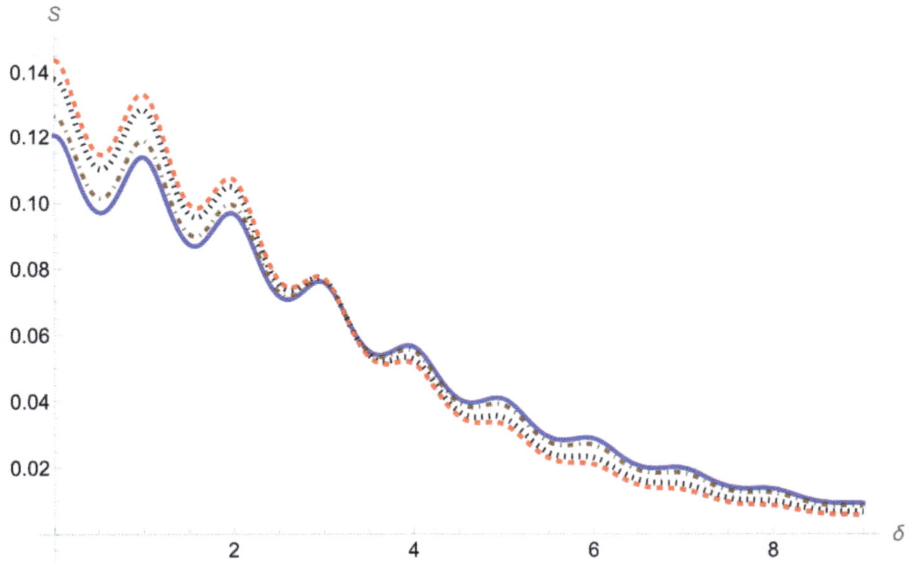

Figure 2.94. Calculated profiles of the Lyman-delta line for the scaled laser amplitude $\varepsilon = 1$ for four different observation angles: $\pi/2$ (solid line), $\pi/3$ (dash-dotted line), $\pi/6$ (dotted line), and 0 (dashed line).

Third, the peaks of the satellites intensities have a monotonically decreasing envelope. This differs from the case of the one-mode laser field.

Now we proceed to studying the angular dependence of the Lyman-7 profiles. Figure 2.97 shows its profiles for the scaled laser amplitude $\varepsilon = 0.0625$ for four

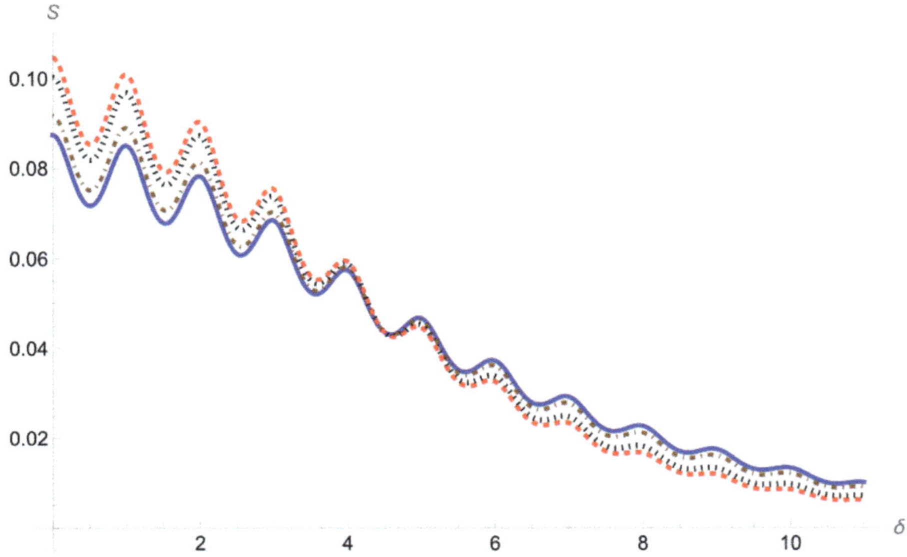

Figure 2.95. Calculated profiles of the Lyman-delta line for the scaled laser amplitude $\varepsilon = 2$ for four different observation angles: $\pi/2$ (solid line), $\pi/3$ (dash-dotted line), $\pi/6$ (dotted line), and 0 (dashed line).

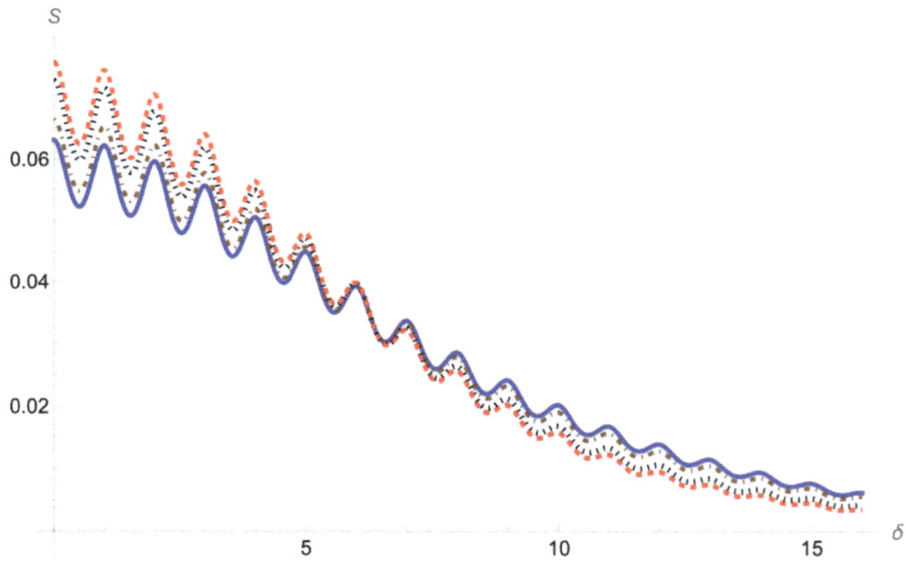

Figure 2.96. Calculated profiles of the Lyman-delta line for the scaled laser amplitude $\varepsilon = 4$ for four different observation angles: $\pi/2$ (solid line), $\pi/3$ (dash-dotted line), $\pi/6$ (dotted line), and 0 (dashed line).

different observation angles: $\pi/2$ (solid line), $\pi/3$ (dash-dotted line), $\pi/6$ (dotted line), and 0 (dashed line).

Figure 2.98 demonstrates the calculated profiles of the Lyman-7 line for the scaled laser amplitude $\varepsilon = 0.125$ for four different observation angles: $\pi/2$ (solid line), $\pi/3$ (dash-dotted line), $\pi/6$ (dotted line), and 0 (dashed line).

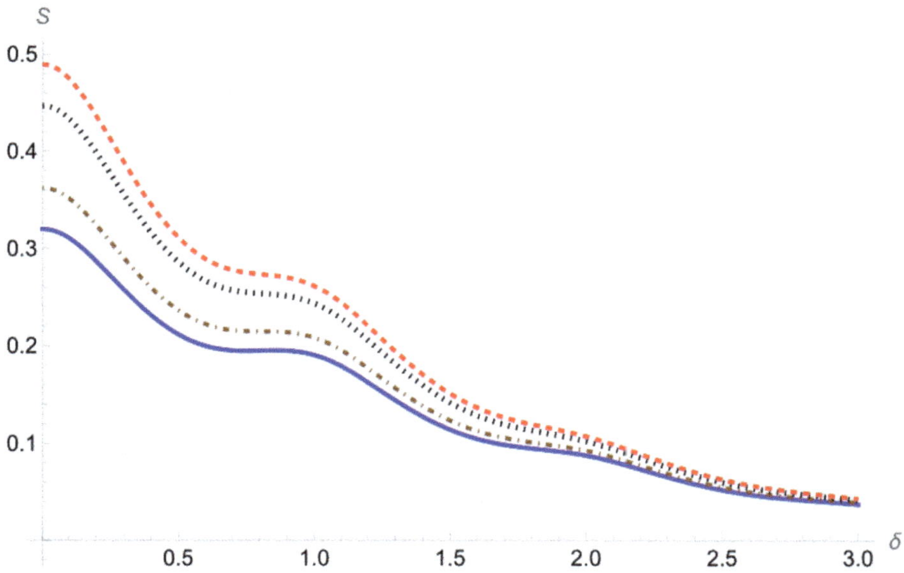

Figure 2.97. Calculated profiles of the Lyman-7 line for the scaled laser amplitude $\varepsilon = 0.0625$ for four different observation angles: $\pi/2$ (solid line), $\pi/3$ (dash-dotted line), $\pi/6$ (dotted line), and 0 (dashed line).

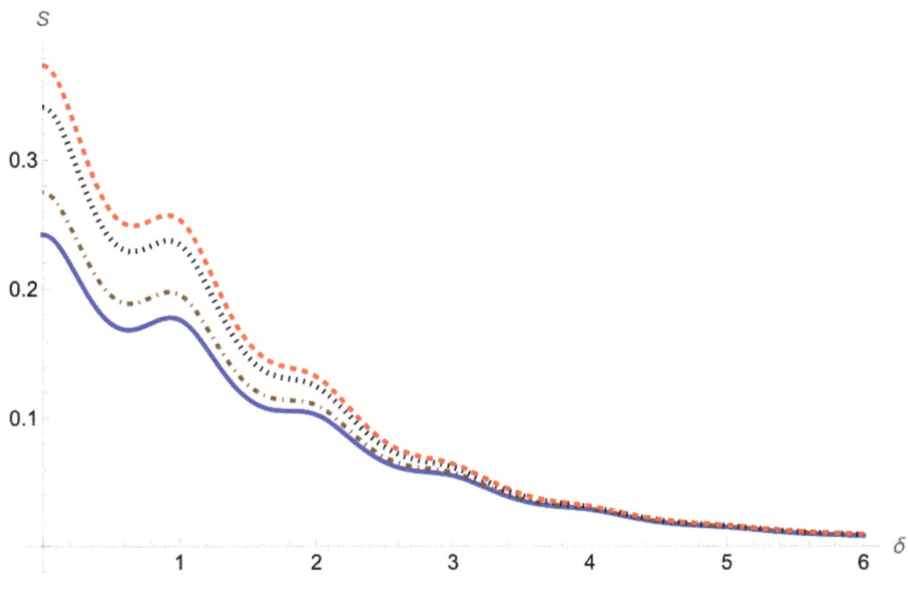

Figure 2.98. Calculated profiles of the Lyman-7 line for the scaled laser amplitude $\varepsilon = 0.125$ for four different observation angles: $\pi/2$ (solid line), $\pi/3$ (dash-dotted line), $\pi/6$ (dotted line), and 0 (dashed line).

Figure 2.99 displays the calculated profiles of the Lyman-7 line for the scaled laser amplitude $\varepsilon = 0.25$ for four different observation angles: $\pi/2$ (solid line), $\pi/3$ (dash-dotted line), $\pi/6$ (dotted line), and 0 (dashed line).

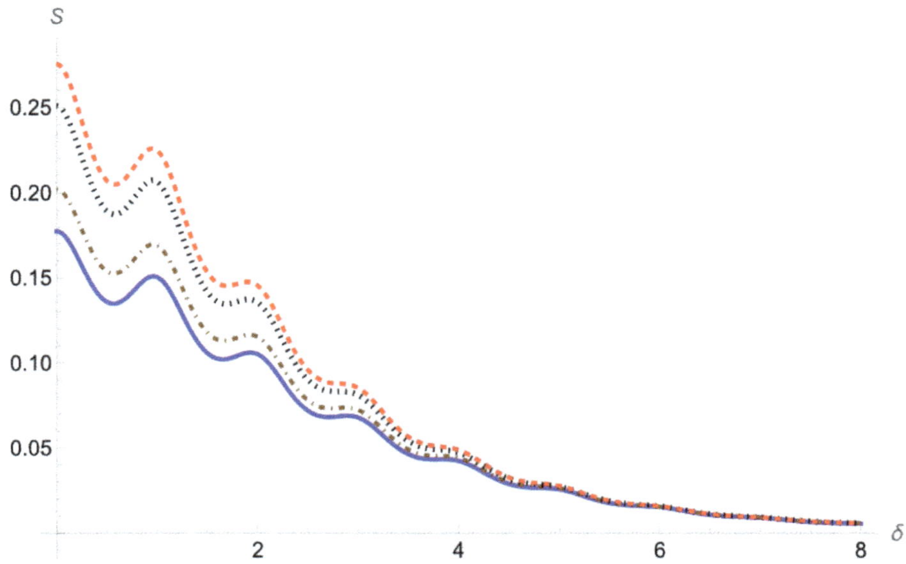

Figure 2.99. Calculated profiles of the Lyman-7 line for the scaled laser amplitude $\varepsilon = 0.25$ for four different observation angles: $\pi/2$ (solid line), $\pi/3$ (dash-dotted line), $\pi/6$ (dotted line), and 0 (dashed line).

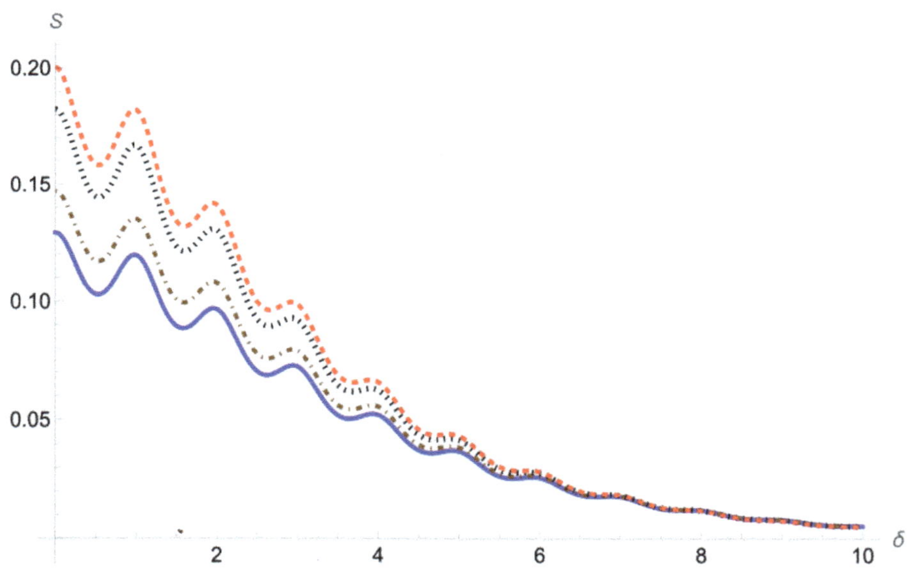

Figure 2.100. Calculated profiles of the Lyman-7 line for the scaled laser amplitude $\varepsilon = 0.5$ for four different observation angles: $\pi/2$ (solid line), $\pi/3$ (dash-dotted line), $\pi/6$ (dotted line), and 0 (dashed line).

Figure 2.100 shows the calculated profiles of the Lyman-7 line for the scaled laser amplitude $\varepsilon = 0.5$ for four different observation angles: $\pi/2$ (solid line), $\pi/3$ (dash-dotted line), $\pi/6$ (dotted line), and 0 (dashed line).

Figure 2.101 demonstrates the calculated profiles of the Lyman-7 line for the scaled laser amplitude $\varepsilon = 1$ for four different observation angles: $\pi/2$ (solid line), $\pi/3$ (dash-dotted line), $\pi/6$ (dotted line), and 0 (dashed line).

Figure 2.102 displays the calculated profiles of the Lyman-7 line for the scaled laser amplitude $\varepsilon = 2$ for four different observation angles: $\pi/2$ (solid line), $\pi/3$ (dash-dotted line), $\pi/6$ (dotted line), and 0 (dashed line).

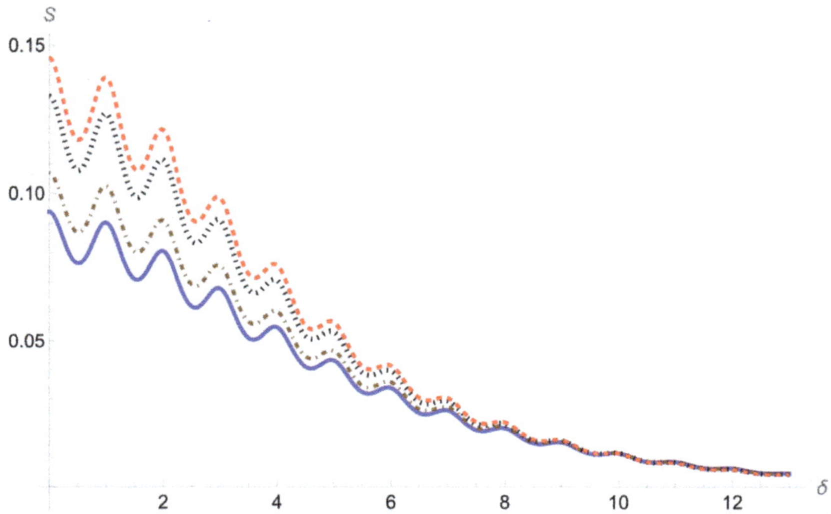

Figure 2.101. Calculated profiles of the Lyman-7 line for the scaled laser amplitude $\varepsilon = 1$ for four different observation angles: $\pi/2$ (solid line), $\pi/3$ (dash-dotted line), $\pi/6$ (dotted line), and 0 (dashed line).

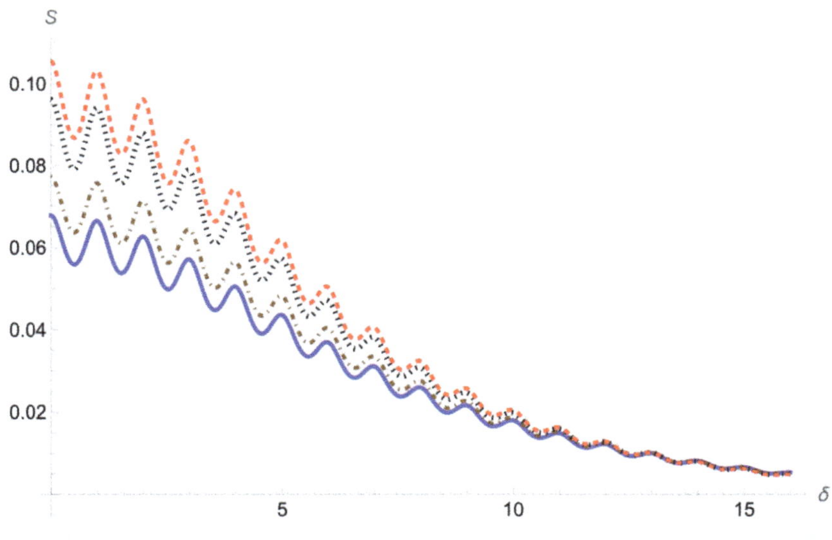

Figure 2.102. Calculated profiles of the Lyman-7 line for the scaled laser amplitude $\varepsilon = 2$ for four different observation angles: $\pi/2$ (solid line), $\pi/3$ (dash-dotted line), $\pi/6$ (dotted line), and 0 (dashed line).

From figures 2.97–2.102 one can see the following. First of all, as the angle of the observation decreases *from π/2 to 0, the half width at half maximum diminishes very significantly. This is a clear* distinction from the corresponding profiles of the Lyman-beta and Lyman delta lines, where such decrease was not so pronounced.

Second, the peaks of the satellites intensities have a monotonically decreasing envelope—just as for the corresponding profiles of the Lyman-beta and Lyman-delta lines. This differs from the case of the one-mode laser field.

Now we proceed to studying the angular dependence of the Lyman-9 profiles. Figure 2.103 shows its profiles for the scaled laser amplitude $\varepsilon = 0.031\ 25$ for four different observation angles: $\pi/2$ (solid line), $\pi/3$ (dash-dotted line), $\pi/6$ (dotted line), and 0 (dashed line).

Figure 2.104 demonstrates the calculated profiles of the Lyman-9 line for the scaled laser amplitude $\varepsilon = 0.0625$ for four different observation angles: $\pi/2$ (solid line), $\pi/3$ (dash-dotted line), $\pi/6$ (dotted line), and 0 (dashed line).

Figure 2.105 displays the calculated profiles of the Lyman-9 line for the scaled laser amplitude $\varepsilon = 0.125$ for four different observation angles: $\pi/2$ (solid line), $\pi/3$ (dash-dotted line), $\pi/6$ (dotted line), and 0 (dashed line).

Figure 2.106 shows the calculated profiles of the Lyman-9 line for the scaled laser amplitude $\varepsilon = 0.25$ for four different observation angles: $\pi/2$ (solid line), $\pi/3$ (dash-dotted line), $\pi/6$ (dotted line), and 0 (dashed line).

Figure 2.107 demonstrates the calculated profiles of the Lyman-9 line for the scaled laser amplitude $\varepsilon = 0.5$ for four different observation angles: $\pi/2$ (solid line), $\pi/3$ (dash-dotted line), $\pi/6$ (dotted line), and 0 (dashed line).

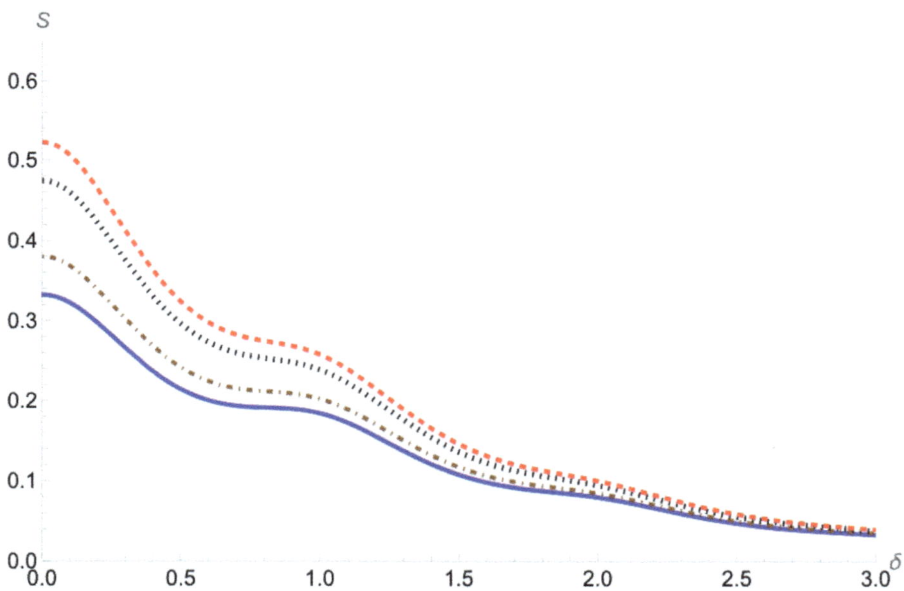

Figure 2.103. Calculated profiles of the Lyman-9 line for the scaled laser amplitude $\varepsilon = 0.031\ 25$ for four different observation angles: $\pi/2$ (solid line), $\pi/3$ (dash-dotted line), $\pi/6$ (dotted line), and 0 (dashed line).

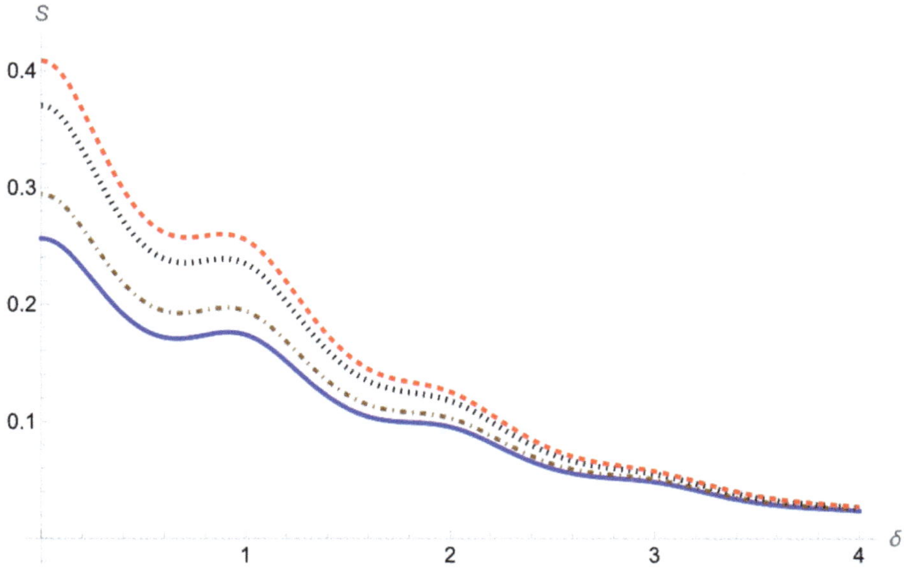

Figure 2.104. Calculated profiles of the Lyman-9 line for the scaled laser amplitude $\varepsilon = 0.0625$ for four different observation angles: $\pi/2$ (solid line), $\pi/3$ (dash-dotted line), $\pi/6$ (dotted line), and 0 (dashed line).

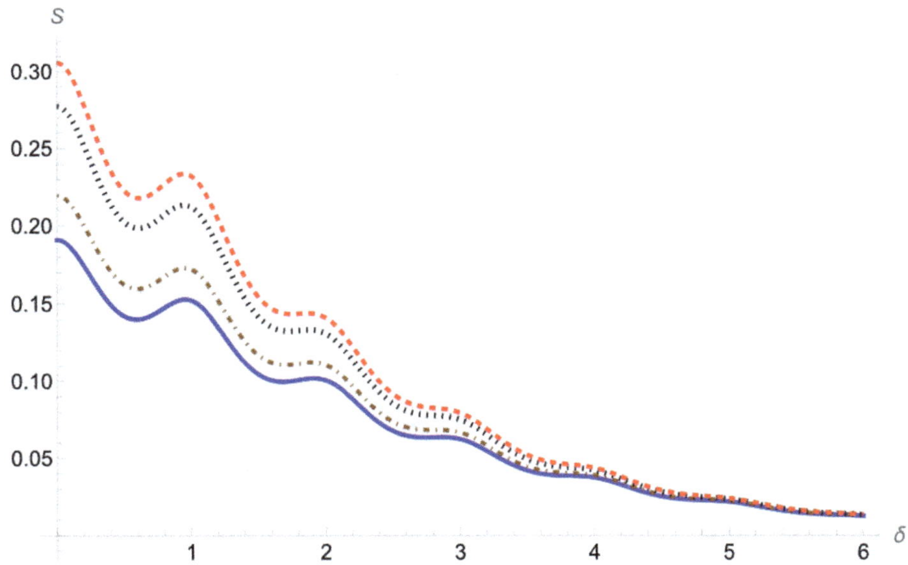

Figure 2.105. Calculated profiles of the Lyman-9 line for the scaled laser amplitude $\varepsilon = 0.125$ for four different observation angles: $\pi/2$ (solid line), $\pi/3$ (dash-dotted line), $\pi/6$ (dotted line), and 0 (dashed line).

Figure 2.108 displays the calculated profiles of the Lyman-9 line for the scaled laser amplitude $\varepsilon = 1$ for four different observation angles: $\pi/2$ (solid line), $\pi/3$ (dash-dotted line), $\pi/6$ (dotted line), and 0 (dashed line).

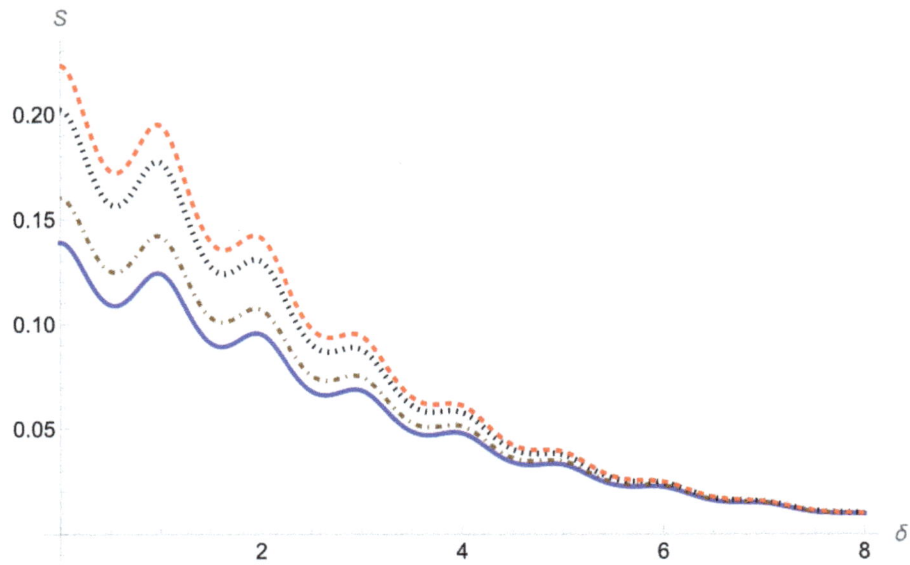

Figure 2.106. Calculated profiles of the Lyman-9 line for the scaled laser amplitude $\varepsilon = 0.25$ for four different observation angles: $\pi/2$ (solid line), $\pi/3$ (dash-dotted line), $\pi/6$ (dotted line), and 0 (dashed line).

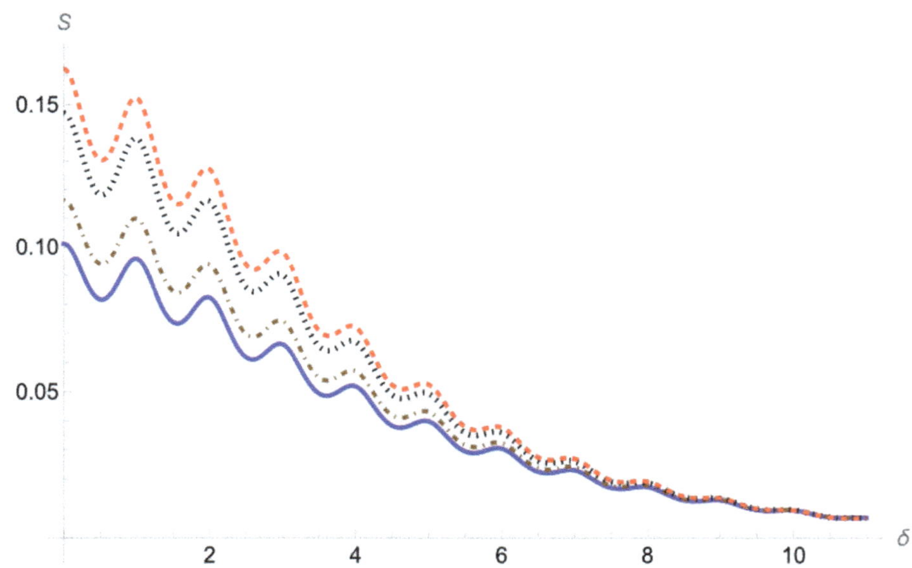

Figure 2.107. Calculated profiles of the Lyman-9 line for the scaled laser amplitude $\varepsilon = 0.5$ for four different observation angles: $\pi/2$ (solid line), $\pi/3$ (dash-dotted line), $\pi/6$ (dotted line), and 0 (dashed line).

From figures 2.103–2.108 one can see the following. The profiles of the Lyman-9 line are very sensitive to the direction of the observation, just as the corresponding profiles of the Lyman-7 line. As the angle of the observation decreases from $\pi/2$ to 0, the half width at half maximum diminishes very significantly. This is a clear

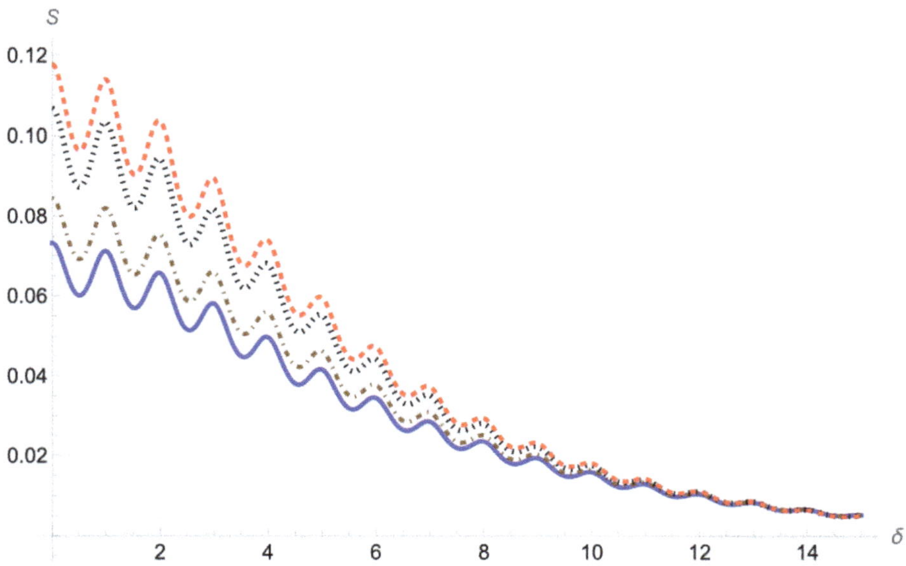

Figure 2.108. Calculated profiles of the Lyman-9 line for the scaled laser amplitude $\varepsilon = 1$ for four different observation angles: $\pi/2$ (solid line), $\pi/3$ (dash-dotted line), $\pi/6$ (dotted line), and 0 (dashed line).

distinction from the corresponding profiles of the Lyman-beta and Lyman delta lines, where such decrease was not so pronounced.

Second, the profiles of the Lyman-9 lines are more sensitive to the amplitude of the laser field compared to the corresponding profiles of the lines Lyman-beta, Lyman-delta, and Lyman-7.

Third, the peaks of the satellites intensities have a monotonically decreasing envelope—just as for the corresponding profiles of the Lyman-beta and Lyman-delta lines. This differs from the case of the one-mode laser field.

Now we proceed to studying the angular dependence of the Lyman-11 profiles. Figure 2.109 shows its profiles for the scaled laser amplitude $\varepsilon = 0.031\ 25$ for four different observation angles: $\pi/2$ (solid line), $\pi/3$ (dash-dotted line), $\pi/6$ (dotted line), and 0 (dashed line).

Figure 2.110 demonstrates the calculated profiles of the Lyman-11 line for the scaled laser amplitude $\varepsilon = 0.0625$ for four different observation angles: $\pi/2$ (solid line), $\pi/3$ (dash-dotted line), $\pi/6$ (dotted line), and 0 (dashed line).

Figure 2.111 displays the calculated profiles of the Lyman-11 line for the scaled laser amplitude $\varepsilon = 0.125$ for four different observation angles: $\pi/2$ (solid line), $\pi/3$ (dash-dotted line), $\pi/6$ (dotted line), and 0 (dashed line).

Figure 2.112 shows the calculated profiles of the Lyman-11 line for the scaled laser amplitude $\varepsilon = 0.25$ for four different observation angles: $\pi/2$ (solid line), $\pi/3$ (dash-dotted line), $\pi/6$ (dotted line), and 0 (dashed line).

Figure 2.113 demonstrates the calculated profiles of the Lyman-11 line for the scaled laser amplitude $\varepsilon = 0.5$ for four different observation angles: $\pi/2$ (solid line), $\pi/3$ (dash-dotted line), $\pi/6$ (dotted line), and 0 (dashed line).

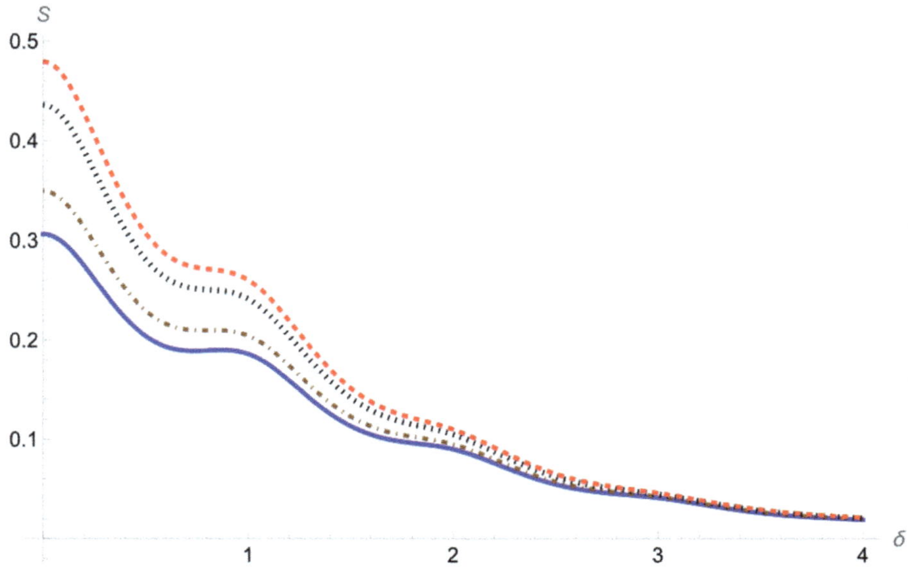

Figure 2.109. Calculated profiles of the Lyman-11 line for the scaled laser amplitude $\varepsilon = 0.031\ 25$ for four different observation angles: $\pi/2$ (solid line), $\pi/3$ (dash-dotted line), $\pi/6$ (dotted line), and 0 (dashed line).

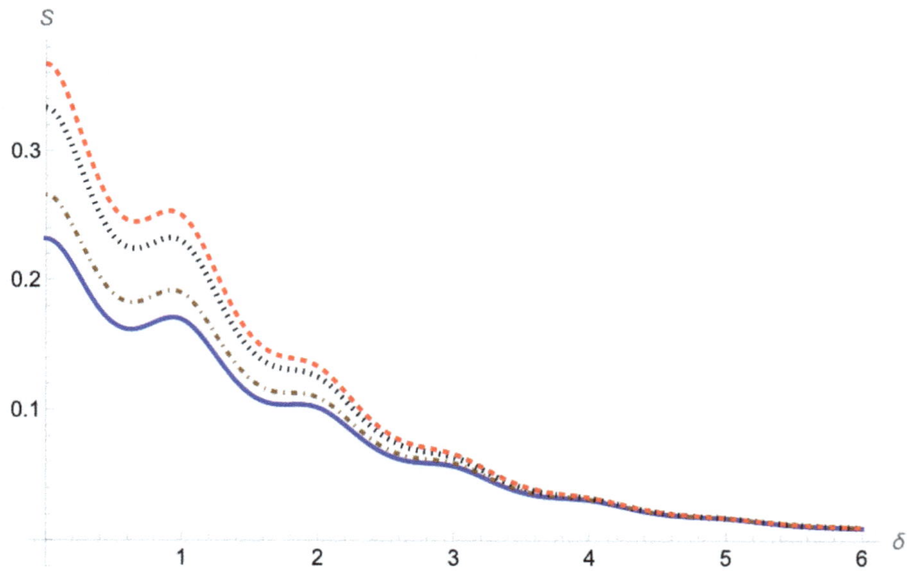

Figure 2.110. Calculated profiles of the Lyman-11 line for the scaled laser amplitude $\varepsilon = 0.0625$ for four different observation angles: $\pi/2$ (solid line), $\pi/3$ (dash-dotted line), $\pi/6$ (dotted line), and 0 (dashed line).

Figure 2.114 demonstrates the calculated profiles of the Lyman-11 line for the scaled laser amplitude $\varepsilon = 1$ for four different observation angles: $\pi/2$ (solid line), $\pi/3$ (dash-dotted line), $\pi/6$ (dotted line), and 0 (dashed line).

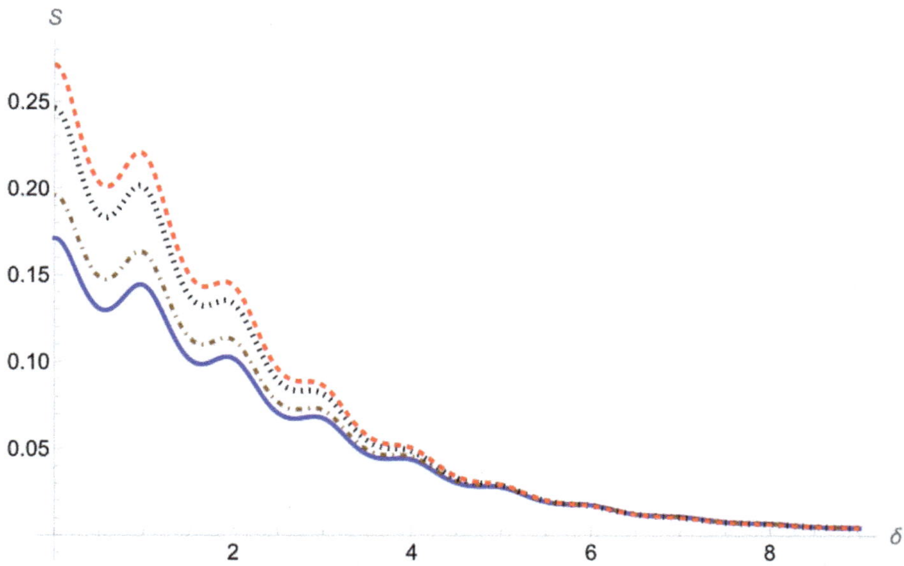

Figure 2.111. Calculated profiles of the Lyman-11 line for the scaled laser amplitude $\varepsilon = 0.125$ for four different observation angles: $\pi/2$ (solid line), $\pi/3$ (dash-dotted line), $\pi/6$ (dotted line), and 0 (dashed line).

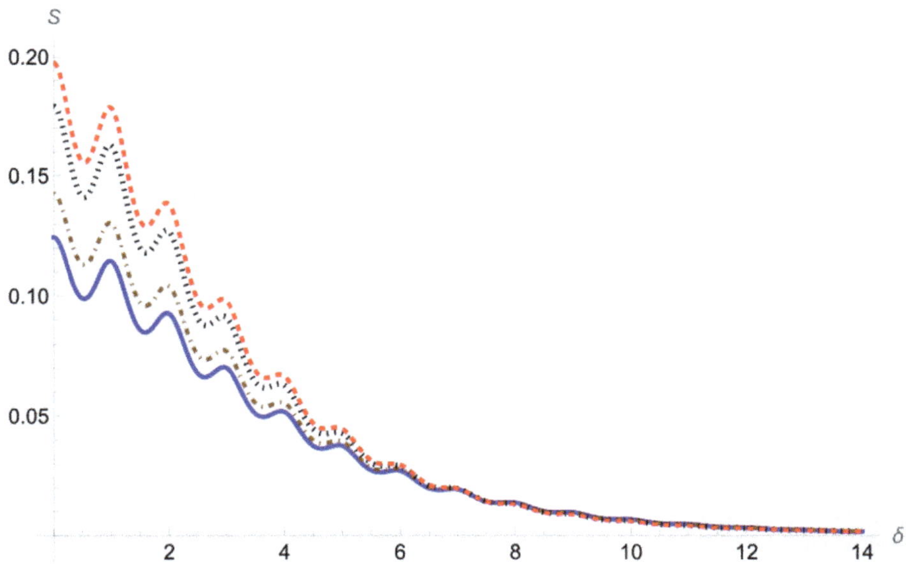

Figure 2.112. Calculated profiles of the Lyman-11 line for the scaled laser amplitude $\varepsilon = 0.25$ for four different observation angles: $\pi/2$ (solid line), $\pi/3$ (dash-dotted line), $\pi/6$ (dotted line), and 0 (dashed line).

From figures 2.109–2.114 one can see the following. The profiles of the Lyman-11 line are even more sensitive to the direction of the observation compared the corresponding profiles of the Lyman-7 and Lyman-9 lines. As the angle of the observation decreases from $\pi/2$ to 0, the half width at half maximum diminishes very

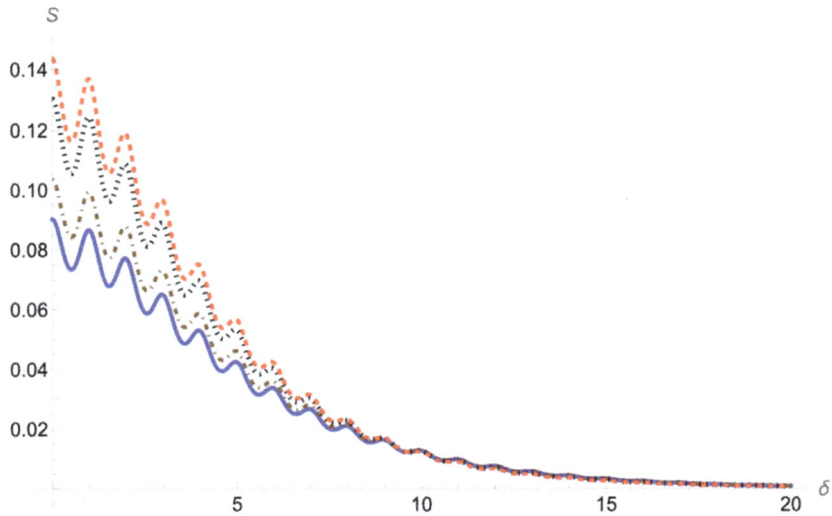

Figure 2.113. Calculated profiles of the Lyman-11 line for the scaled laser amplitude $\varepsilon = 0.5$ for four different observation angles: $\pi/2$ (solid line), $\pi/3$ (dash-dotted line), $\pi/6$ (dotted line), and 0 (dashed line).

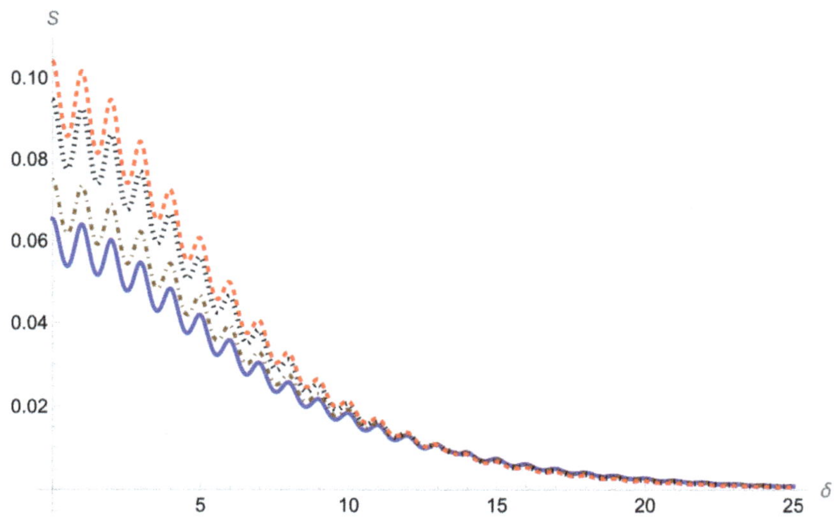

Figure 2.114. Calculated profiles of the Lyman-11 line for the scaled laser amplitude $\varepsilon = 1$ for four different observation angles: $\pi/2$ (solid line), $\pi/3$ (dash-dotted line), $\pi/6$ (dotted line), and 0 (dashed line).

significantly. This is a clear distinction from the corresponding profiles of the Lyman-beta and Lyman delta lines, where such decrease was not so pronounced.

Second, the profiles of the Lyman-11 lines are more sensitive to the amplitude of the laser field compared to the corresponding profiles of the lines Lyman-beta, Lyman-delta, Lyman-7, and Lyman-9.

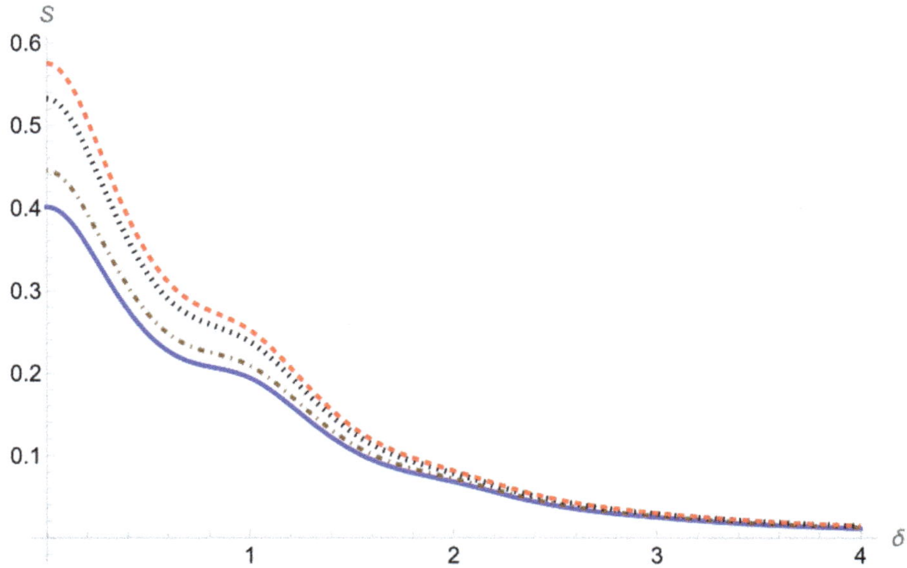

Figure 2.115. Calculated profiles of the Balmer-beta line for the scaled laser amplitude $\varepsilon = 0.125$ for four different observation angles: $\pi/2$ (solid line), $\pi/3$ (dash-dotted line), $\pi/6$ (dotted line), and 0 (dashed line).

Third, again the peaks of the satellites intensities have a monotonically decreasing envelope—just as for the corresponding profiles of the Lyman-beta, Lyman-delta, Lyman-7, and Lyman-9 lines. This differs from the case of the one-mode laser field.

Now we proceed to studying the angular dependence of the Balmer-beta profiles. Figure 2.115 shows its profiles for the scaled laser amplitude $\varepsilon = 0.125$ for four different observation angles: $\pi/2$ (solid line), $\pi/3$ (dash-dotted line), $\pi/6$ (dotted line), and 0 (dashed line).

Figure 2.116 demonstrates the calculated profiles of the Balmer-beta line for the scaled laser amplitude $\varepsilon = 0.25$ for four different observation angles: $\pi/2$ (solid line), $\pi/3$ (dash-dotted line), $\pi/6$ (dotted line), and 0 (dashed line).

Figure 2.117 displays the calculated profiles of the Balmer-beta line for the scaled laser amplitude $\varepsilon = 0.5$ for four different observation angles: $\pi/2$ (solid line), $\pi/3$ (dash-dotted line), $\pi/6$ (dotted line), and 0 (dashed line).

Figure 2.118 shows the calculated profiles of the Balmer-beta line for the scaled laser amplitude $\varepsilon = 1$ for four different observation angles: $\pi/2$ (solid line), $\pi/3$ (dash-dotted line), $\pi/6$ (dotted line), and 0 (dashed line).

Figure 2.119 demonstrates the calculated profiles of the Balmer-beta line for the scaled laser amplitude $\varepsilon = 2$ for four different observation angles: $\pi/2$ (solid line), $\pi/3$ (dash-dotted line), $\pi/6$ (dotted line), and 0 (dashed line).

Figure 2.120 displays the calculated profiles of the Balmer-beta line for the scaled laser amplitude $\varepsilon = 4$ for four different observation angles: $\pi/2$ (solid line), $\pi/3$ (dash-dotted line), $\pi/6$ (dotted line), and 0 (dashed line).

From figures 2.115–2.120 one can see the following. The profiles of the Balmer-beta line are noticeably sensitive to the direction of the observation compared the

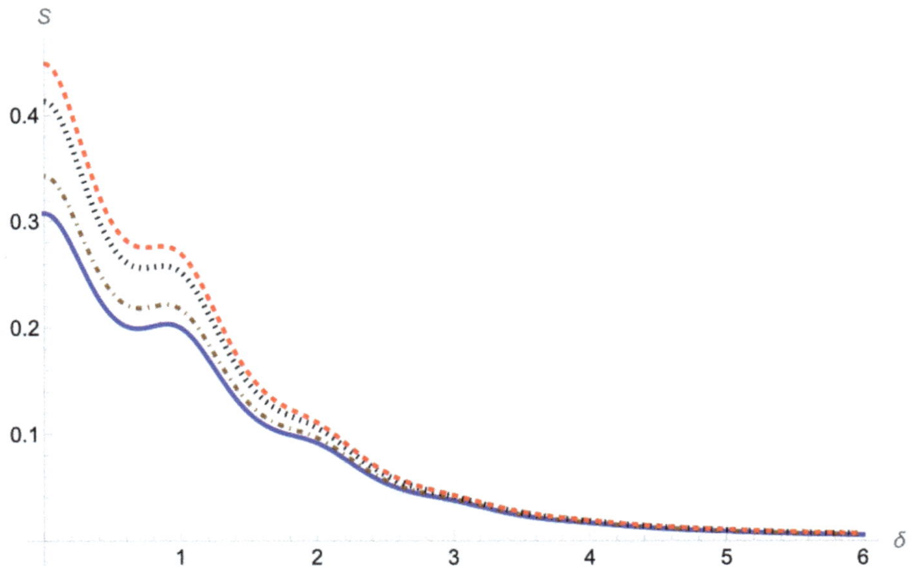

Figure 2.116. Calculated profiles of the Balmer-beta line for the scaled laser amplitude $\varepsilon = 0.25$ for four different observation angles: $\pi/2$ (solid line), $\pi/3$ (dash-dotted line), $\pi/6$ (dotted line), and 0 (dashed line).

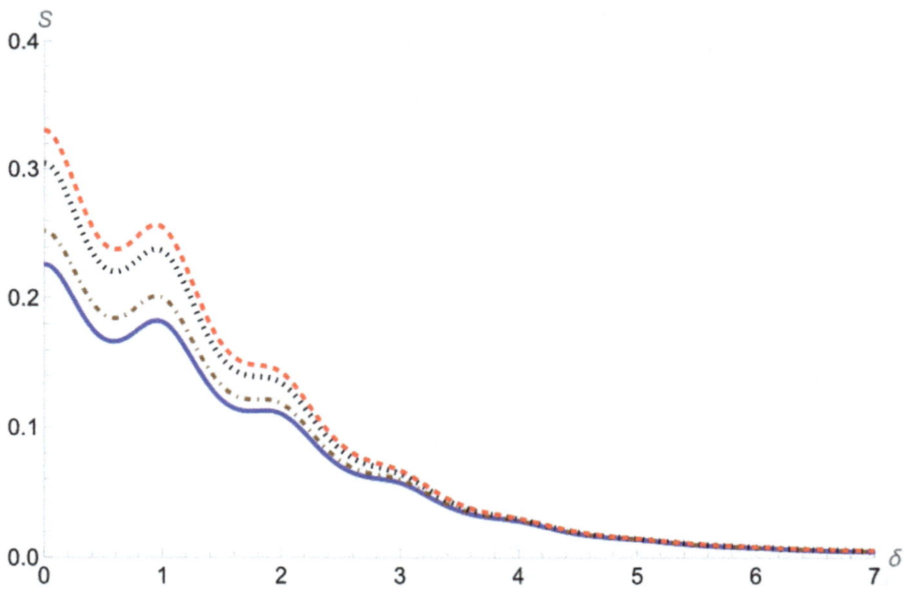

Figure 2.117. Calculated profiles of the Balmer-beta line for the scaled laser amplitude $\varepsilon = 0.5$ for four different observation angles: $\pi/2$ (solid line), $\pi/3$ (dash-dotted line), $\pi/6$ (dotted line), and 0 (dashed line).

corresponding profiles of the Lyman-beta and Lyman-delta lines. As the angle of the observation decreases from $\pi/2$ to 0, the half width at half maximum diminishes quite clearly.

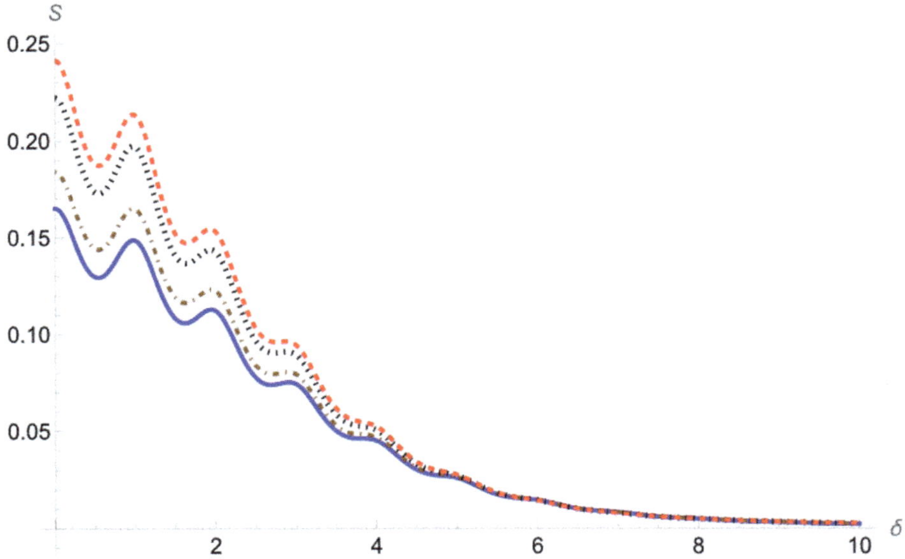

Figure 2.118. Calculated profiles of the Balmer-beta line for the scaled laser amplitude $\varepsilon = 1$ for four different observation angles: $\pi/2$ (solid line), $\pi/3$ (dash-dotted line), $\pi/6$ (dotted line), and 0 (dashed line).

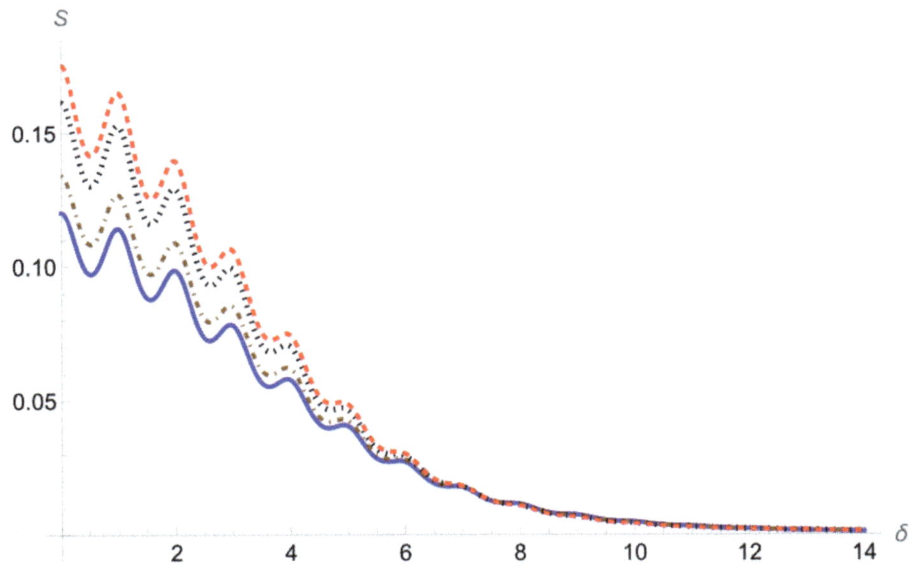

Figure 2.119. Calculated profiles of the Balmer-beta line for the scaled laser amplitude $\varepsilon = 2$ for four different observation angles: $\pi/2$ (solid line), $\pi/3$ (dash-dotted line), $\pi/6$ (dotted line), and 0 (dashed line).

Second, the profiles of the Balmer-beta lines are more sensitive to the amplitude of the laser field compared to the corresponding profiles of the Lyman-beta line

Third, again the peaks of the satellites intensities have a monotonically decreasing envelope—just as for the corresponding profiles of the Lyman lines studies above.

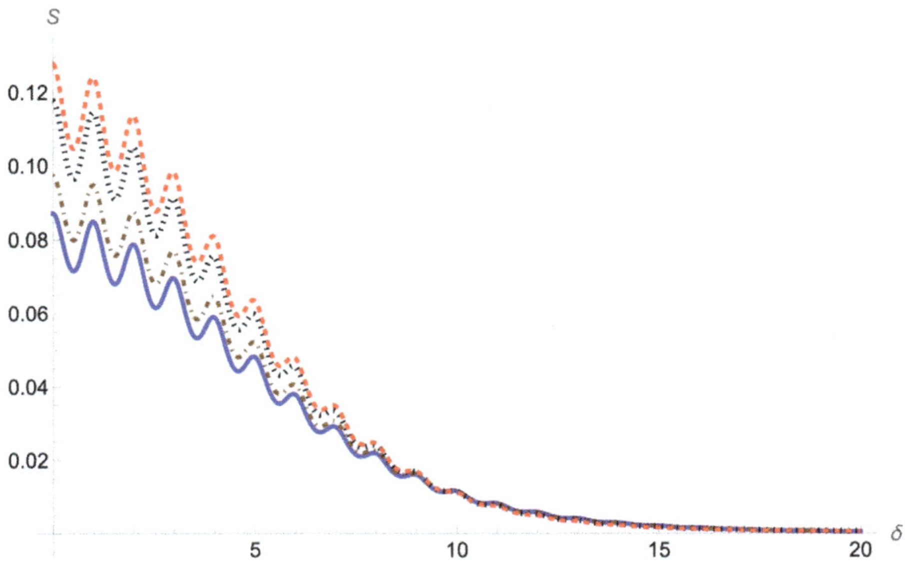

Figure 2.120. Calculated profiles of the Balmer-beta line for the scaled laser amplitude $\varepsilon = 4$ for four different observation angles: $\pi/2$ (solid line), $\pi/3$ (dash-dotted line), $\pi/6$ (dotted line), and 0 (dashed line).

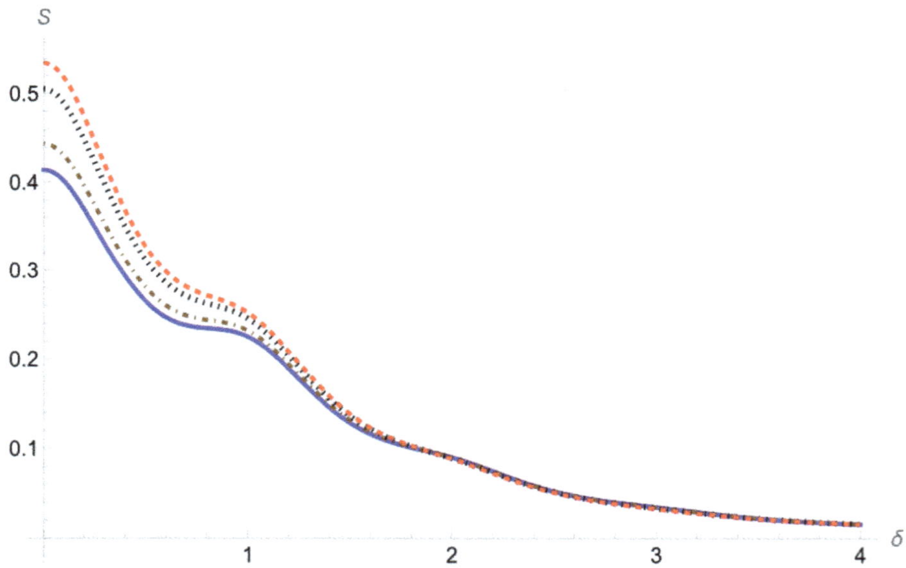

Figure 2.121. Calculated profiles of the Balmer-delta line for the scaled laser amplitude $\varepsilon = 0.0625$ for four different observation angles: $\pi/2$ (solid line), $\pi/3$ (dash-dotted line), $\pi/6$ (dotted line), and 0 (dashed line).

Now we proceed to studying the angular dependence of the Balmer-delta profiles. Figure 2.121 shows its profiles for the scaled laser amplitude $\varepsilon = 0.0625$ for four different observation angles: $\pi/2$ (solid line), $\pi/3$ (dash-dotted line), $\pi/6$ (dotted line), and 0 (dashed line).

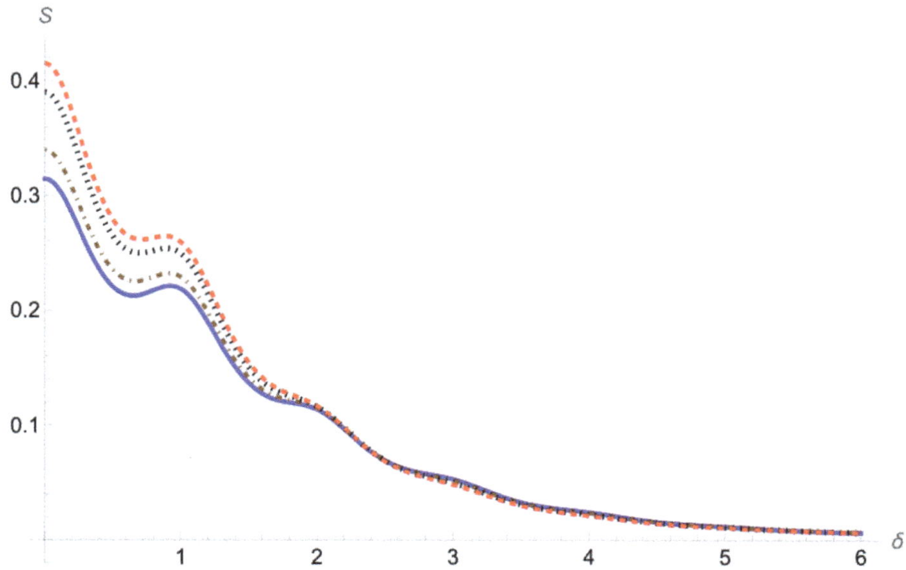

Figure 2.122. Calculated profiles of the Balmer-delta line for the scaled laser amplitude ε = 0.125 for four different observation angles: π/2 (solid line), π/3 (dash-dotted line), π/6 (dotted line), and 0 (dashed line).

Figure 2.122 demonstrates the calculated profiles of the Balmer-delta line for the scaled laser amplitude ε = 0.125 for four different observation angles: π/2 (solid line), π/3 (dash-dotted line), π/6 (dotted line), and 0 (dashed line).

Figure 2.123 displays the calculated profiles of the Balmer-delta line for the scaled laser amplitude ε = 0.25 for four different observation angles: π/2 (solid line), π/3 (dash-dotted line), π/6 (dotted line), and 0 (dashed line).

Figure 2.124 shows the calculated profiles of the Balmer-delta line for the scaled laser amplitude ε = 0.5 for four different observation angles: π/2 (solid line), π/3 (dash-dotted line), π/6 (dotted line), and 0 (dashed line).

Figure 2.125 demonstrates the calculated profiles of the Balmer-delta line for the scaled laser amplitude ε = 1 for four different observation angles: π/2 (solid line), π/3 (dash-dotted line), π/6 (dotted line), and 0 (dashed line).

Figure 2.125 displays the calculated profiles of the Balmer-delta line for the scaled laser amplitude ε = 2 for four different observation angles: π/2 (solid line), π/3 (dash-dotted line), π/6 (dotted line), and 0 (dashed line).

From figures 2.121–2.126 one can see the following. First, quite unexpectedly, the profiles of the Balmer-delta line are less sensitive to the direction of the observation compared the corresponding profiles of the Balmer-beta line. As the angle of the observation decreases from π/2 to 0, the half width at half maximum diminishes only slightly.

Second, the profiles of the Balmer-delta lines are more sensitive to the amplitude of the laser field compared to the corresponding profiles of the Balmer-beta line

Third, again the peaks of the satellites intensities have a monotonically decreasing envelope—just as for the corresponding profiles of the Balmer-beta line.

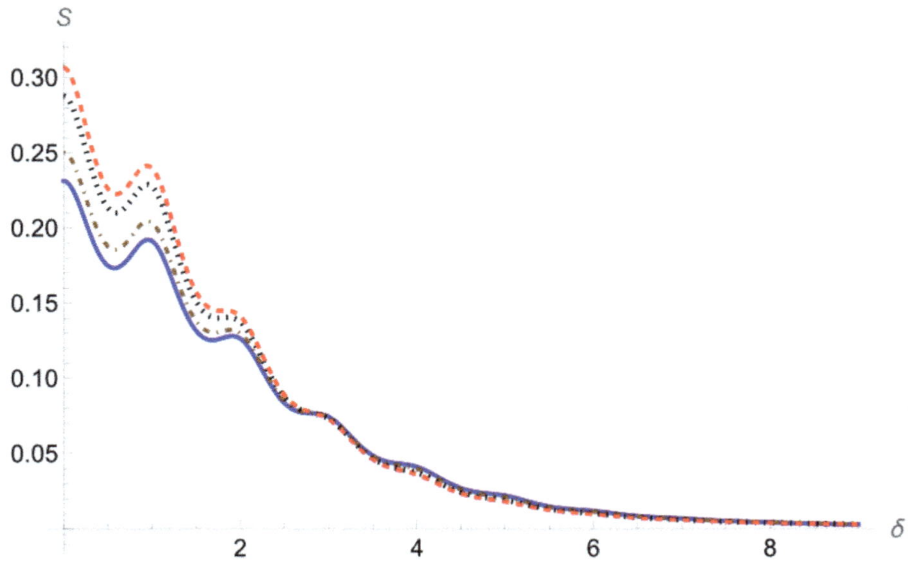

Figure 2.123. Calculated profiles of the Balmer-delta line for the scaled laser amplitude $\varepsilon = 0.25$ for four different observation angles: $\pi/2$ (solid line), $\pi/3$ (dash-dotted line), $\pi/6$ (dotted line), and 0 (dashed line).

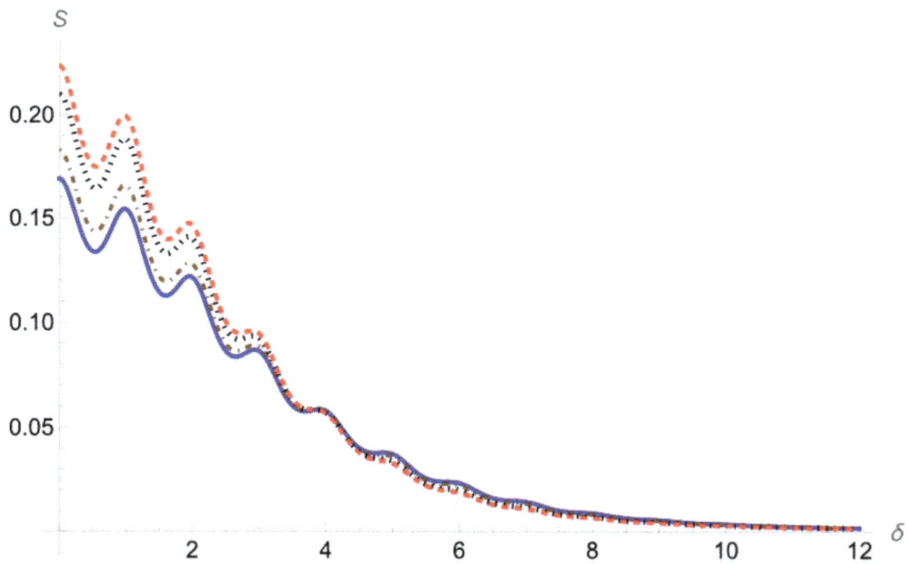

Figure 2.124. Calculated profiles of the Balmer-delta line for the scaled laser amplitude $\varepsilon = 0.5$ for four different observation angles: $\pi/2$ (solid line), $\pi/3$ (dash-dotted line), $\pi/6$ (dotted line), and 0 (dashed line).

Summarizing the section 2.3, we could provide the following application of the results for diagnosing the multi-mode laser field in plasmas, where due to nonlinear processes, the amplitude of the electromagnetic field at the laser frequency differs from the amplitude of the incoming laser field. The one-dimensional multi-mode

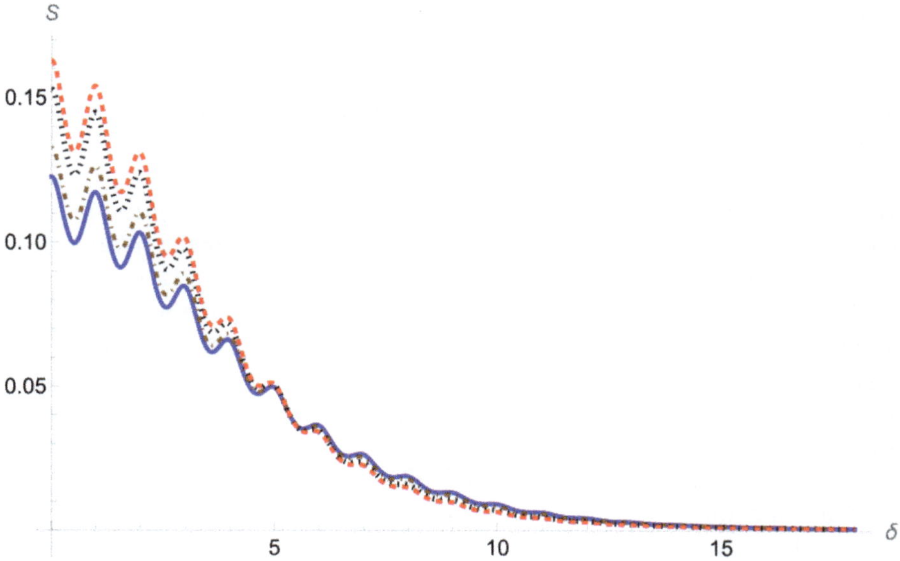

Figure 2.125. Calculated profiles of the Balmer-delta line for the scaled laser amplitude $\varepsilon = 1$ for four different observation angles: $\pi/2$ (solid line), $\pi/3$ (dash-dotted line), $\pi/6$ (dotted line), and 0 (dashed line).

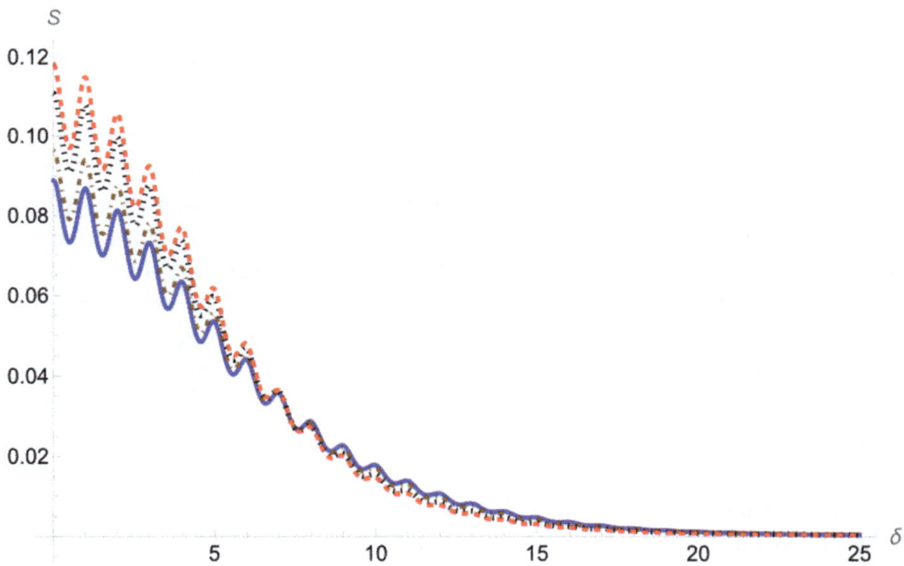

Figure 2.126. Calculated profiles of the Balmer-delta line for the scaled laser amplitude $\varepsilon = 2$ for four different observation angles: $\pi/2$ (solid line), $\pi/3$ (dash-dotted line), $\pi/6$ (dotted line), and 0 (dashed line).

monochromatic electric field can create numerous satellites of hydrogenic spectral lines —just as the two-mode laser field. For the experimental determination of whether the observed profiles correspond to the multi-mode field or to the two-mode field, the best choice is the line Lyman-9 (if possible). Indeed, if the observed profile of the Lyman-9 line, the envelope of the satellites peaks exhibits the secondary maximum, then the laser field is two-mode rather than multi-mode.

As for the other spectral lines, it would be necessary to measure profiles of at least two spectral lines: at the same laser amplitude, they have different halfwidths of the envelopes of the satellites peaks—this would allow the experimental determination of the amplitude of the multi-mode laser field despite the presence of other line broadening mechanisms in plasmas. For enhancing the reliability of the determination of the amplitude of the electromagnetic field at the laser frequency inside the plasma, it is a good idea to compare the profiles of the same spectral line observed at different angles—thus taking the advantage of the directional effects described above.

Below we present an example of the practical implementation of diagnosing the one-dimensional multi-mode field of Langmuir oscillation in the plasma produced by the Phoenix plasma radiation source [5]. The plasma radiation source is a pulsed power Z-pinch employed for producing an intense radiation in the x-ray range. By using crystal spectrometers, profiles of several Lyman lines of the hydrogenic Al XIII have been observed and analyzed in detail.

Specifically, the Z-pinch power was rapidly delivered to a wire load that consisted of 16 aluminum wires. The wires had a length of 2 cm and diameter of 0.08 cm. They were assembled in a circular array. Due to this geometry, the resulting plasma were stretched along the direction of the wires.

The primary diagnostic tools were two Johann spectrometers and one flat crystal spectrometer.

The flat crystal spectrometer was designed to simultaneously observe the profiles of the Lyman-delta and Lyman-epsilon lines of Al XIII in one state of polarization. The most important information was obtained by using the Johann spectrometers. They were designed in the following way—based on the polarization effect of the Brewster angle of incidence on the crystal. One of them obtained either Lyman-beta or Lyman-gamma line profiles polarized parallel to the symmetry axis of the plasma, while the other spectrometer obtained the corresponding profiles polarized perpendicular to the symmetry axis of the plasma. This was the first time that the profiles of the same x-ray spectral line were observed in two mutually orthogonal polarizations simultaneously (to the best of our knowledge).

Figure 2.127 shows the experimental profiles of the Lyman-gamma line in the polarization parallel and perpendicular to the plasma axis of symmetry. It is seen that the profiles in the two polarizations differ from each other.

The comprehensive analysis of all experimental profiles, taking into account all broadening mechanisms, led to the following conclusion. The primary broadening mechanism was due to the quasi-one-dimensional multi-mode electric field of Langmuir waves, that is, waves at the plasma electron frequency $\omega_{pe} = (4\pi e^2 N_e / m_e)$. This agreed with the plasma theory suggesting that these Langmuir waves were caused by runaway electrons in the discharge.

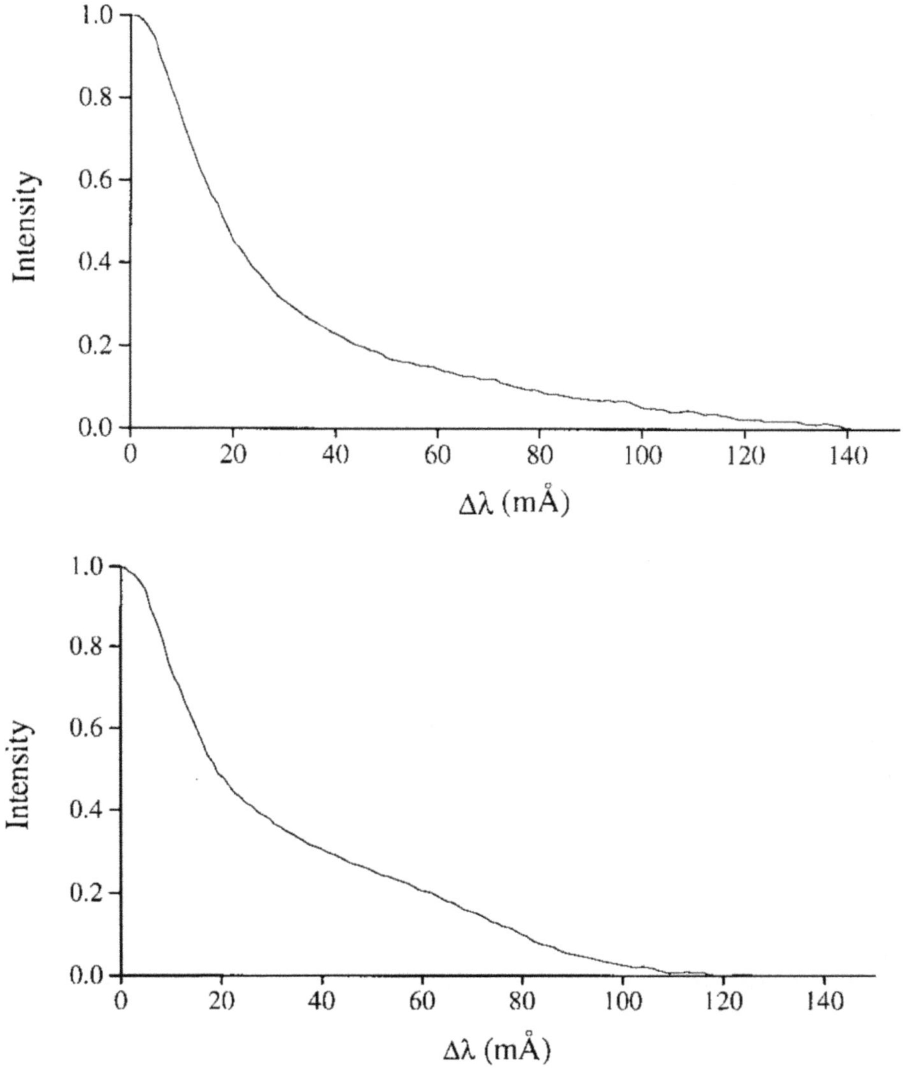

Figure 2.127. Experimental profile of the Lyman-gamma line of hydrogenic Al XIII observed in the polarization parallel to the plasma symmetry axis (the top part) and in the polariztion perpendicular to the plasma syummetry axis (the bottom part) at the Phoenix plasma radiation source. Copyright 1998 IEEE. Reprinted, with permission, from [5].

2.4 Satellites under the elliptically-polarized monochromatic electric field

The effect of an elliptically-polarized laser field on hydrogenic spectral lines was analyzed, for instance, in works [6, 7]. The interaction of an elliptically-polarized field with a hydrogenic atom/ion of the nuclear charge Z can be written in the form

$$V(t) = x\xi\varepsilon_0 \sin \omega t + z\varepsilon_0 \cos \omega t, \tag{2.16}$$

where

$$\varepsilon_0 = E_0(1 + \xi^2)^{1/2}. \tag{2.17}$$

The Schrödinger equation is:

$$i\,\partial\Psi/\partial t = [H_a + V(t)]\Psi. \tag{2.18}$$

The electric field rotates with the frequency

$$\omega_1(t) = \xi\omega/(\cos^2\omega t + \xi^2 \text{sinn}^2\omega t), \tag{2.19}$$

where $\omega_1(t)$ is the time derivative of the angular coordinate $\phi(t)$ of the electric field vector. In the coordinate system rotating with the same frequency, the Schrödinger equation can be reformulated as follows:

$$i\,\partial\Phi/\partial t = [H_a + zE(t) - \omega_1(t)L_y]\Phi, \tag{2.20}$$

where

$$\Phi(t) = \exp[i\varphi(t)L_y]\Psi(t), \tag{2.21}$$

and

$$E(t) = \varepsilon_0(1 - k^2\sin^2\omega t)^{1/2}, \quad k = 1 - \xi^2. \tag{2.22}$$

In equations (2.20) and (2.21), L_y is the y-component of the angular momentum operator.

The relative significance of the 'magnetic' ($-\omega_1(t)L_y$) and electric interactions is controlled by the ratio

$$X(t) = 2Z\,\omega_1(t)/[3nE(t)]. \tag{2.23}$$

The most physically interesting is the case where $X^2(t) \ll 1$, leading to a resonance effect.

For $n = 2$, the instantaneous eigenvalues of the Hamiltonian $H(t)$ are as follows (according to papers [8, 9]):

$$\begin{aligned} W_1(t) &= -\,W_4(t) = 3E(t)[1 + X^2(t)]^{1/2}/Z \equiv W(t), \\ W_2(t) &\equiv W_3(t) \equiv 0. \end{aligned} \tag{2.24}$$

For the case where $X^2(t) \ll 1$, the corresponding instantaneous eigenfunctions can be written as follows:

$$\psi_{1\text{inst}} - \psi_1 = \psi_{4\text{inst}} - \psi_4 = -(\psi_1 + \psi_4)X^2/4 + i(\psi_2 + \psi_3)X/2, \tag{2.25}$$

$$\psi_{2\text{inst}} = i(\psi_1 + \psi_4)X/2 + \psi_2(1 + X^2/4) - \psi_3 X^2/2, \tag{2.26}$$

$$\psi_{3\text{inst}} = i(\psi_1 + \psi_4)X/2 + \psi_3(1 - X^2/4). \tag{2.27}$$

In equations (2.25), (2.26), and (2.27), ψ_j are the standard parabolic wave functions that diagonalize the Hamiltonian $H_a + zE(t)$:

$$\psi_1 = |100\rangle, \psi_2 = |100\rangle, \psi_3 = |00-1\rangle, \psi_4 = |010\rangle. \tag{2.28}$$

By using the instantaneous eigenfunctions and eigenvalues as the adiabatic basis, there was obtained the analytical solution of equation (2.20) in the formalism of the quasienergy states (the states introduced in papers [10, 11]). First, for the hypothetical situation, where $X(t) = 0$, the separations Q between these quasienergy states were expressed through the time averaged instantaneous splitting $W_n(t)$ in the mutually perpendicular electric and 'magnetic' fields

$$\begin{aligned} Q &= \langle W_n(t)\rangle + 2r\omega \quad (r = 0, \pm 1, \ldots), \\ \langle W_n(t)\rangle &\approx 3n\varepsilon_0 b_0/Z + Z\xi^2\omega^2 g_0/6n\varepsilon_0. \end{aligned} \tag{2.29}$$

Here and below, the formulas containing the principal quantum number n represent the more general results that are valid for any n.

In reality, the non-zero value of $X(t)$ causes transitions between the quasienergy states. If $Q \approx 2q\omega$, where $q = 1, 2, 3, \ldots$, there occurs multiphoton resonances between many harmonics of the quasienergy states, the resonances being simultaneously due to *all* harmonics of $X(t)$. The resonance condition can be finally represented in the form

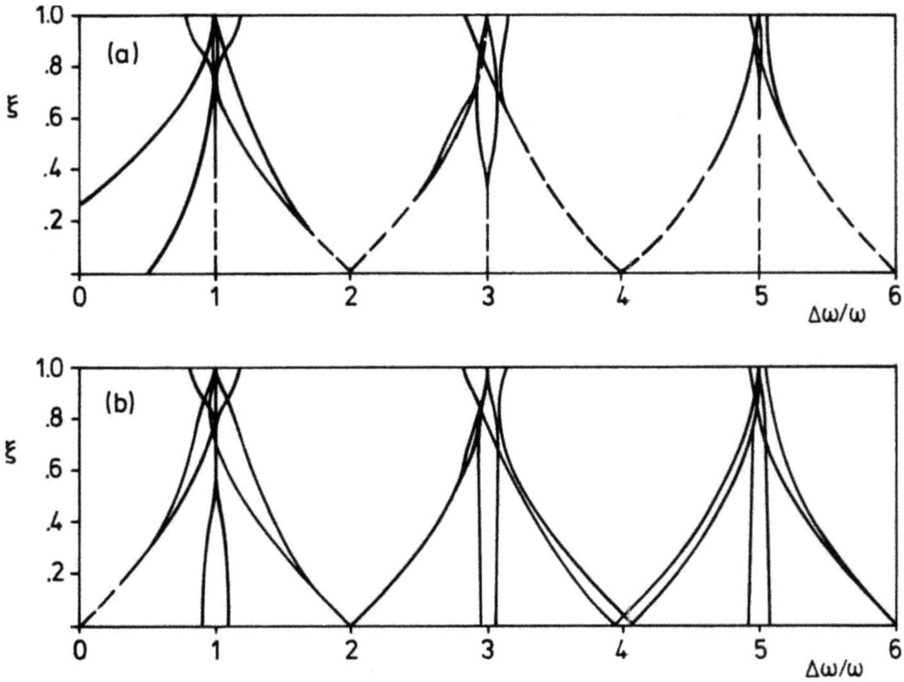

Figure 2.128. The dependence of the spectrum of the Lyman-alpha line on the ellipticity degree ξ for $3E_0/(Z\omega) = 5.5$ in two polarizations. Intense spectral components are shown as bands (as double lines). Their position corresponds to the median of the band; their intensity is proportional to the width of the band: (a) x-polarization; (b) z-polarization. Reprinted from [2], copyright (1995), with permission from Springer Nature.

$$\langle W_n(t) \rangle = 2s\omega, \ s = 1, 2, 3, \ldots, \tag{2.30}$$

where Δ is a detuning ($|\Delta| \ll \omega$).

The solution obtained in the resonance approximation allowed calculating the profiles of the Lyman-alpha line. Figure 2.128 presents the dependence of the Lyman-alpha spectrum on the ellipticity degree ξ for $3E_0/(Z\omega) = 5.5$ in two polarizations. In the vicinity of $\xi \approx 0.6$, there occurs a four-photon resonance ($s = 2$), leading to a dramatic modifications of the spectrum, such as abrupt changes of intensities and the appearance of new spectral components.

References

[1] Blochinzew D I 1933 *Phys. Z. Sow. Union* **4** 501

[2] Oks E 1995 *Plasma Spectroscopy: The Influence of Microwave and Laser Fields* (Berlin: Springer)

[3] Renner O, Peyrusse O, Sondhauss P and Förster E 2000 *J. Phys. B: At. Mol. Opt. Phys.* **33** L151

[4] Lifshitz E V 1968 *Sov. Phys. JETP* **26** 570

[5] Weinheimer J, Oks E, Clothiaux E J, Schulz A and Svidzinski V 1998 *IEEE Trans. Plasma Sci.* **26** 1239

[6] Oks E and Gavrilenko V P 1983 *Opt. Commun.* **46** 205

[7] Volod'ko D A, Gavrilenko V P and Oks E 1987 *18th Int. Conf. on Phenomena in Ionized Gases (Swansea, UK)* p 604

[8] Ishimura T 1967 *J. Phys. Soc. Jpn.* **23** 422

[9] Lisitsa V S 1971 *Opt. Spectrosc.* **31** 468

[10] Zel'dovich J B 1967 *Sov. Phys. JETP* **24** 1006

[11] Ritus V I 1967 *Sov. Phys. JETP* **24** 1041

Chapter 3

Polarization and directional effects in the radiation of satellites of non-hydrogenic spectral lines from plasmas

Satellites of the spectral lines of helium and helium-like ions, as well as satellites of the spectral lines of lithium and lithium-like are most frequently used in practice (apart from hydrogenic spectral lines) for the spectroscopic diagnostics of the Oscillatory Electric Fields (OEFs) in various plasmas. Of course, this tool has to be 'calibrated' by the corresponding theory.

The early corresponding theories, based on the standard Dirac perturbation theory, were presented in papers [1, 2]. A more advanced theory—the adiabatic perturbation theory valid beyond the range of theories from papers [1, 2]—was developed in paper [3]. The results from papers [1–3] were also represented in books [4, 5].

In all these calculations there were considered three levels 0, 1, 2, such that the two closely lying levels 1 and 2 are coupled by a dipole matrix element. Levels 1 and 2 are separated by the distance Δ in the frequency scale (here and below atomic units are used). The radiative transition from 1 to a distant level 0 is dipole-forbidden, while the radiative transition from 2 to 0 is dipole-allowed.

The satellites are polarized. This can be utilized for obtaining the information about the anisotropy of the distribution of the QEFs or for measuring the angle between the transmission axis of the polarizer and the vector $\boldsymbol{E}(t) = \boldsymbol{E}_0 \cos \omega t$. In works [3, 4], we chose the y-axis along the direction of the observation. The two possible orthogonal directions of the transmission axis of the polarizer—in the xz-plane—define the directions of the z- and x-axes. The vector $\mathbf{E}_0$, located in the xz-plane, constitutes the angle γ with the z-axis—measuring this angle is one of the diagnostic purposes.

doi:10.1088/978-0-7503-6285-6ch3

The satellites intensities, corresponding to the z- and x-directions of the polarizer, are denoted $S_{\pm}{}^{(z)}$ and $S_{\pm}{}^{(x)}$, respectively. The plus or minus in the subscripts correspond to the satellite at the location $+\omega$ or $-\omega$ (counted from the unperturbed position of the spectral line), respectively. Those intensities are as follows:

$$\begin{aligned} S_{\mp}^{(x)} &= \sum_{m,m',m''} \sigma_{\mp}(|x_{20}|^2 \sin^2\gamma + |z_{20}|^2 \cos^2\gamma), \\ S_{\mp}^{(x)} &= \sum_{m,m',m''} \sigma_{\mp}(|x_{20}|^2 \cos^2\gamma + |z_{20}|^2 \sin^2\gamma). \end{aligned} \tag{3.1}$$

In equation (3.1),

$$\sigma_{\pm} \approx \frac{1}{(\Delta \pm \omega)^2}\left[\left(\frac{E_0 z_{12}}{2}\right)^2 \pm \left(\frac{E_0 z_{12}}{2}\right)^4 \frac{\Delta^3 \mp 7\Delta^2\omega + 3\Delta\omega^2 \mp \omega^3}{\omega(\Delta^2 - \omega^2)^2}\right]. \tag{3.2}$$

The ratio of the intensities of the same satellite in two orthogonal polarizations is as follows:

$$\begin{aligned} S_{\mp}^{(x)}/S_{\mp}^{(x)} &= [1 + f_{\mp}(E_0)\cot^2\gamma]/[\cot^2\gamma + f_{\mp}(E_0)], \\ f_{\mp}(E_0) &= \left(\sum_{m,m',m'',} \sigma_{\mp}|z_{20}|^2\right) \Big/ \left(\sum_{m,m',m''} \sigma_{\mp}|x_{20}|^2\right). \end{aligned} \tag{3.3}$$

This diagnostic method was practically implemented for the first time in paper [6].

Another interesting situation is where the OEF is in resonance with the perturbed separation between levels 1 and 2. The most pronounced resonance effects are when the perturbed separation between levels 1 and 2 is equal to the odd number of quanta of the OEF (see section 5.2 of book [4]).

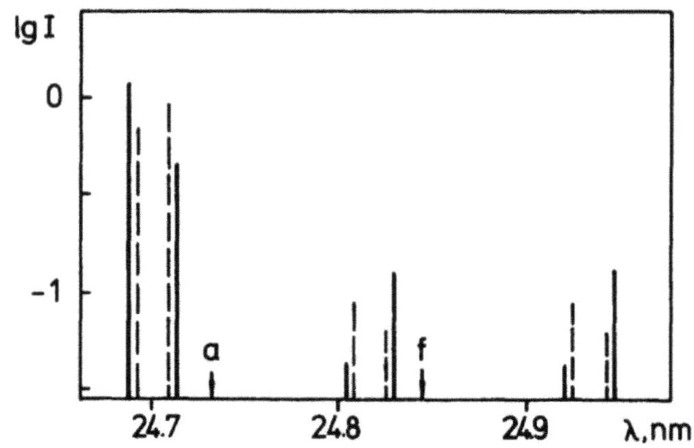

Figure 3.1. The spectrum of the radiative transitions from the singlet levels 3^1D and 3^1P to the singlet level 2^1P in the hydrogenic carbon (C V) in the conditions of the 3-quantum resonance with the CO_2 laser of the amplitude $E_0 = 40$ MV cm^{-1}. The π-components are shown by solid lines, while the σ-components are shown by the dashed lines. The unperturbed positions of the allowed (a) and forbidden (f) lines are indicated by the arrows. Reprinted from [4], copyright (1995), with permission from Springer Nature.

As an illustrative example, we considered the spectral line of the helium-like carbon (C V) in the field of the CO_2 laser. Specifically, the closely lying singlet levels 3^1D and 3^1P were chosen as levels 1 and 2, respectively. The radiative transitions were considered to occur to the singlet level 2^1S (thus playing the role of level 0). The separation between the unperturbed levels 3^1D and 3^1P, $\Delta/(2\pi c)$ is 1846 cm^{-1}. The frequency of the CO_2 laser is $\omega/(2\pi c) = 943$ cm^{-1}. The laser field increases the separation between the levels 3^1D and 3^1P and can bring them to the 3-quantum resonance with the frequency of this laser—the range of the laser amplitude $\varepsilon_0 \approx$ (30–50) MV cm^{-1}. The result is the splitting of both the main spectral line and its satellites. The splitting depends on the laser amplitude. The π-components split differently from the σ-components—thus, the polarization effect, as illustrated by figure 3.1.

For further reading we refer the reader to works [7–11].

References

[1] Baranger M and Mozer B 1961 *Phys. Rev.* **123** 25

[2] Cooper W S and Ringler H 1969 *Phys. Rev.* **179** 226

[3] Oks E and Gavrilenko V P 1983 *Sov. Tech. Phys. Lett.* **9** 111

[4] Oks E 1995 *Plasma Spectroscopy: The Influence of Microwave and Laser Fields* (Berlin: Springer)

[5] Oks E 2017 *Diagnostics of Laboratory and Astrophysical Plasmas Using Spectral Lines of One-, Two-, and Three-Electron Systems* (Hackensack, NJ: World Scientific)

[6] Brizhinev M P, Gavrilenko V P, Egorov S V, Eremin B G, Kostrov A V, Oks E and Shagiev Y M 1983 *Sov. Phys. JETP* **58** 517

[7] Perelman N F and Mosyak A A 1989 *Sov. Phys. JETP* **69** 700

[8] Gavrilenko V P and Oks E 1982 *Proc. Int. Conf. on Plasma Physics (Göteborg, Sweden)* p 353

[9] Gavrilenko V P and Oks E 1983 *Sov. J. Quantum Electron.* **13** 1269

[10] Gavrilenko V P and Oks E 1989 *Opt. Commun.* **69** 384

[11] Gavrilenko V P and Oks E 1986 *Proc. 13th Summer School and Int. Symp. on Physics of Ionized Gases (Sibenik, Yugoslavia)* p 393

Chapter 4

Polarization and directional effects in the intra-Stark spectroscopy of plasmas

The research area of the intra-Doppler spectroscopy was known for well over half-a-century. The root-cause of that research area were nonlinear optical phenomena causing some depressions to show up at certain locations within the Doppler profile of a spectral line. The intra-Stark spectroscopy, which is younger than the intra-Doppler spectroscopy, was called so because of its analogy with the intra-Doppler spectroscopy: it deals with some depressions at certain locations within the quasistatic Stark profile of spectral lines in plasmas—the depressions caused by a different type of nonlinear optical phenomena compared to the intra-Doppler spectroscopy. The term 'intra-Stark spectroscopy' was introduced for the first time in papers [1, 2].

The intra-Stark spectroscopy deals with the Langmuir-wave-caused 'dips' in the profiles of spectral lines emitted from plasmas. Here the term 'dip' is utilized just for brevity: in reality it refers to highly-localized structures consisting of a local minimum of the intensity and two adjacent 'bumps' (peaks). The physics behind this emergent phenomenon has to do with multifrequency nonlinear dynamic resonances coupling the quasimonochromatic electric field $\mathbf{E}(t)$ at the characteristic frequency ω in plasmas, the quasistatic electric field $\mathbf{F}$ in plasmas, and the radiating atom or ion. Specifically, there occurs the dynamic resonance between the frequency ω or its harmonics and the Stark splitting of hydrogenic spectral lines. In distinction to the intra-Doppler spectroscopy, this dynamic resonance has a multifrequency nature despite the applied electric field being quasimonochromatic—see paper [3]. For the detailed theory of the intra-Stark spectroscopy, we refer the reader to books [4, 5].

In the corresponding experimental studies, the quasimonochromatic electric field was represented by a Langmuir wave, which is why the observed localized structures in the spectral line profiles we called the L-dips, though readers should keep in mind

doi:10.1088/978-0-7503-6285-6ch4

that these structures frequently consist of the primary minimum ('the dip') and two surrounding bumps.

The practical importance of studying the L-dips is three-fold:

(1) from the experimental positions of the L-dips, the electron density N_e in plasmas can be deduced much more accurately than the value of N_e determined experimentally from the broadening of the spectral lines; the L-dip-based passive spectroscopic method allows measuring N_e with the same high accuracy as active spectroscopic methods, such as, for instance, the more complicated experimental Thompson scattering, as demonstrated in paper [6];

(2) they provide the only one non-perturbative method for deducing the amplitude of the Langmuir waves plasmas;

(3) they exhibit the rich physics of the nonrelativistic and relativistic laser–plasma interaction.

Book [4] and later reviews [7–9] exemplify the experimental studies of the L-dips —the studies focused on spectroscopic diagnostics of a variety of plasmas. In these experiments and observations covering about 10 orders of magnitude range of the electron densities, the Langmuir-wave-caused 'dip' structures were detected, consistently identified, and utilized for plasma diagnostics.

Recent theoretical results on the intra-Stark spectroscopy were published in papers [10–12]. They contain three different new methods, based on the intra-Stark spectroscopy for the experimental determination of super-strong magnetic fields, the fields being in the GigaGauss or multi-GigaGauss range. Such fields are produced at the surface of the relativistic critical density during relativistic laser–plasma interactions [13–22].

Below are some details of the physics underlying the phenomenon of the Langmuir-wave-caused structures in the spectral line profiles. The total electric field is

$$\mathbf{E}(t) = \mathbf{F} + \mathbf{E}_0 \cos(\omega t). \tag{4.1}$$

The quasistatic electric field $\mathbf{F}$ generally represents the sum of the ion microfield (that is, of its quasistatic part) and the low-frequency electrostatic plasma turbulence (examples being the Bernstein modes, or the lower hybrid waves, or the ion acoustic waves). For almost all possible mutual orientations of the two components of the total electric field $\mathbf{E}(t)$, the latter *librates*. Its frequency spectrum is $u\omega$, where $u = 1, 2, 3, \ldots$.

If one were to disregard the librating behavior of the total electric field, then there would be the splitting of the energy levels of the radiator (which is the hydrogenic atom or ion) in 2n−1 Stark sublevels, whose separation (in atomic units) is:

$$\Omega = 3nF_{\text{eff}}/(2Z_r). \tag{4.2}$$

In equation (4.2), n is the principal quantum number and F_{eff} is the averaged absolute value of $\mathbf{E}(t)$—averaged over the libration period:

$$F_{\text{eff}} = \langle|\mathbf{E}(t)|\rangle. \tag{4.3}$$

The Stark sublevels are labeled by q, which is the difference of two parabolic quantum numbers:

$$q = n_1 - n_2. \tag{4.4}$$

For describing physical systems containing the radiating atom (or ion) and the monochromatic field, in papers [23, 24] there was introduced the concept of quasienergy states. For our case under consideration, the quasienergies Q are

$$Q = \Omega + v\omega, \quad v = 0, \pm 1, \pm 2, \pm 3, \ldots \tag{4.5}$$

Now we allow for the time-dependent part of the librating field $\mathbf{E}(t)$, whose frequency spectrum is $u\omega$, where $u = 1, 2, 3, \ldots$. In the case, where

$$\Omega = u\omega, \quad u = 1, 2, 3, \ldots, \tag{4.6}$$

numerous resonances between all quasienergy states of the quasienergies $Q = \Omega + v\omega$ and the harmonics of the librating field take place. This means that the resonances are *multifrequency* and *multiquantum* (involving many quanta of the Langmuir field). This leads to the superposition of the quasienergy harmonics, originating from each of the $2n-1$ atomic Stark substates: all quasienergy states become degenerate.

Further, there is a coupling of the Stark sublevel labeled by a particular value of q with the sublevels labeled by $q-1$ and $q+1$ (the coupling being by the dipole matrix elements), an additional splitting of all quasienergy harmonics takes place. It is the generalized Rabi splitting (the generalization being for the situation of the multi-frequency multiquantum resonances). This generalized Rabi splitting has a non-linear dependence on the amplitude E_0 of the Langmuir wave.

The above resonances correspond to certain resonance values of F_{eff} meeting the condition (4.6), which is why these resonances translate into the bump–dip–bump structures within the quasistatic part of the spectral line profile at specific distances $\Delta\lambda^{\mathrm{dip}}(\omega)$ from the unperturbed position of the spectral line, where $\omega = (4\pi e^2 N_e/m_e)^{1/2}$ of the frequency of the Langmuir wave. Consequently, the positions of these structures are well-defined functions of the electron density (see, e.g., books [4, 5])

$$\Delta\lambda^{\mathrm{dip}} = aN_e^{1/2} + bN_e^{3/4}. \tag{4.7}$$

In equation (4.7), a and b are controlled by the charges of the radiating and perturbing ions, as well as by quantum numbers. This is why from the observed Langmuir 'dips' one can deduce the electron density very accurately. Then from the width of the Langmuir dips one can determine experimentally the amplitude E_0 of the Langmuir wave. This kind of diagnostics was successfully performed in a great variety of laboratory plasmas, as well as in solar plasma—as reviewed in works [4, 7–9].

The Langmuir dips have polarization features, as first revealed in paper [25]. The polarization analysis allows obtaining an additional diagnostic information: namely, about the directionality of the Langmuir waves. The first experimental implementation of the polarization analysis of the Langmuir dips was done in paper [26]. Here are some details.

The experiments were performed at the plasma machine 'Dimpol' [26], which was the plasma trap of the magnetic mirror type. The initial plasma was produced by a Penning-type discharge in the stationary magnetic field. Then a high amplitude ac magnetic field was imposed, creating the configuration of the opposing magnetic fields in the cylindrical geometry described by the coordinates z, r and ϕ.

Polarized profiles of the hydrogen Balmer-alpha line were obtained by introducing a polarizer into the optical system. This enables recording separate z- and ϕ-polarized profiles in the transverse observation, as well as separate r- and ϕ-profiles in the longitudinal observation.

Figure 4.1 shows the experimental profiles of the hydrogen Balmer-alpha line with the polarizer axis in the r-position (a) and in the ϕ-position (b) in the

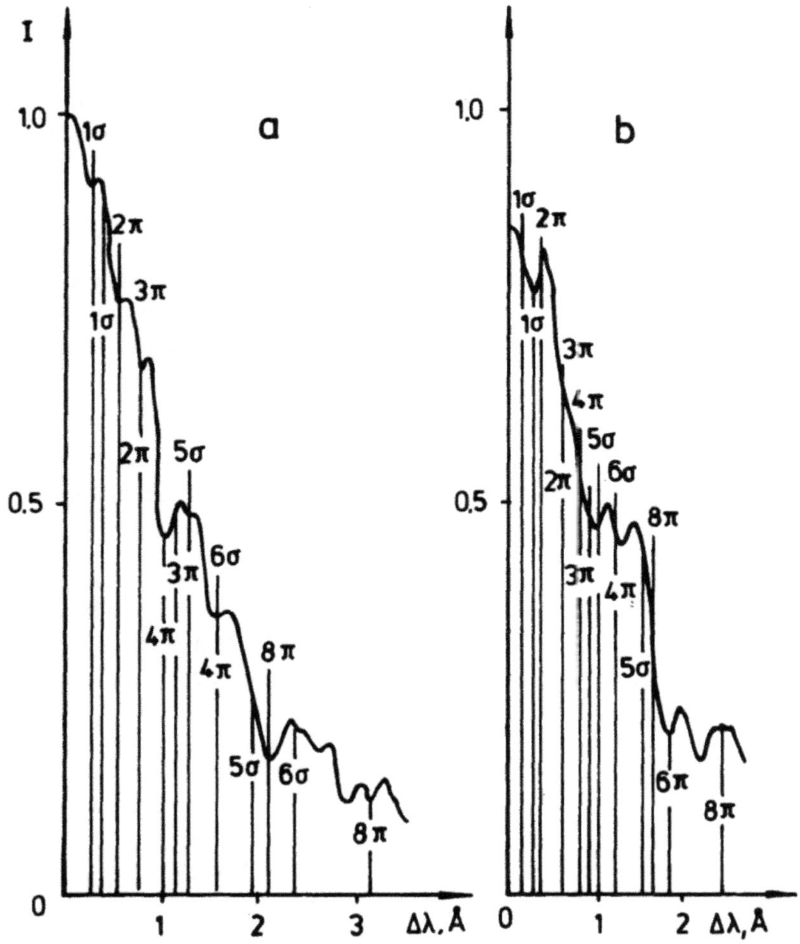

Figure 4.1. Experimental profiles of the hydrogen Balmer-alpha line with the polarizer axis in the r-position (a) and in the ϕ-position (b) in the longitudinal observation in the cylindrical geometry of the plasma machine 'Dimpol'. The segments of the vertical lines point out the theoretically expected positions of the Langmuir dips. For each vertical segment it is indicated the Stark component, to which profile the particular Langmuir dip belongs. Reprinted from [4], copyright (1995), with permission from Springer Nature.

longitudinal observation. The segments of the vertical lines point out the theoretically expected positions of the Langmuir dips. For each vertical segment is indicated the Stark component, to which profile the particular Langmuir dip belongs.

It should be noted that when two bump–dip–bump structures are in close proximity to each other, typically only one bump between the two dips is observed (the two adjacent bumps merged into one). This distorts the width of these two adjacent dips. Therefore, one should analyze the width of the dips in the two polarizations for the most isolated dips. In the experimental profiles from figure 4.1, those are the dips at the following distances from the unperturbed wavelength λ_0 of the spectral line: λ_p. $2\lambda_p$, $3\lambda_p$, and $4\lambda_p$, where

$$\lambda_p = \omega\lambda_0{}^2/(2\pi c). \tag{4.8}$$

The authors of paper [26] performed the statistical comparison of the width of these dips in the z- and ϕ-polarization in the transverse observation and did not find any noticeable difference between the widths of these dips in the two polarizations. This meant:

$$\langle E_z{}^2\rangle \approx \langle E_\varphi{}^2\rangle, \tag{4.9}$$

where $\langle E_z{}^2\rangle$ and $\langle E_\phi{}^2\rangle$ are the average values of the squares of the corresponding projections of the electric field of the Langmuir waves. But the statistical comparison of the width of these dips in the r- and ϕ-polarization in the longitudinal observation revealed that

$$(\Delta\lambda_{1/2}^{\text{dip}})_{r\pi} > (\Delta\lambda_{1/2}^{\text{dip}})_{\varphi\pi},\ (\Delta\lambda_{1/2}^{\text{dip}})_{r\sigma} < (\Delta\lambda_{1/2}^{\text{dip}})_{\varphi\sigma}. \tag{4.10}$$

According to the theoretical results from paper [25], this meant that

$$\langle E_r{}^2\rangle < \langle E_\varphi{}^2\rangle. \tag{4.11}$$

The results expressed in equations (4.9) and (4.11) clearly indicated that the angular distribution of the Langmuir waves had the form of the oblate ellipsoid, whose axis of the symmetry was along r. For further details on the intra-Stark spectroscopy see review article [27].

References

[1] Gavrilenko V P and Oks E 1987 *Sov. Phys. J. Plasma Phys.* **13** 22

[2] Oks E, Böddeker S and Kunze, HJ H-J 1991 *Phys. Rev.* A **44** 8338

[3] Gavrilenko V P and Oks E 1981 *Sov. Phys. JETP* **53** 1122

[4] Oks E 1995 *Plasma Spectroscopy: The Influence of Microwave and Laser Fields* (Berlin: Springer)

[5] Oks E 2017 *Diagnostics of Laboratory and Astrophysical Plasmas Using Spectral Lines of One-, Two-, and Three-Electron Systems* (Hackensack, NJ: World Scientific)

[6] Oks E, Böddeker S and Kunze H-J 1991 *Phys. Rev.* A **44** 8338

[7] Dalimier E, Oks E and Renner O 2014) *Atoms* **2** 178

[8] Dalimier E, Ya. A, Faenov , Oks E, Angelo P, Pikuz T A, Fukuda Y *et al* 2017 *J. Phys.: Conf. Ser.* **810** 012004

[9] Dalimier E, Oks E and Renner O 2017 *AIP Conf. Proc.* **1811** 190003

[10] Dalimier E and Oks E 2018 *Atoms* **6** 60

[11] Oks E, Dalimier E and Angelo P 2019 *Spectrochim. Acta* B **157** 1

[12] Dalimier E, Oks E and Angelo P 2020 *Int. Rev. At. Mol. Phys.* **11** 1

[13] Belyaev V S, Krainov V P, Lisitsa V S and Matafonov A P 2008 *Phys.-Uspekhi* **51** 793

[14] Belyaev V S and Matafonov A P 2011 *Femtosecond-Scale Optics* ed A Andreev (Shanghai: InTech) 70 p 87

[16] Tatarakis M, Gopal A, Watts I, Beg F N, Dangor A E, Krushelnik K *et al* 2022 *Phys. Plasmas* **9** 2244

[17] Tatarakis M, Watts I, Beg F N, Clark E L, Dangor A E, Gopal A *et al* 2002 *Nature* **415** 280

[18] Kato S, Nakamura T, Mima K, Sentoku Y, Nagatomo H and Owadano Y 2004 *J. Plasma Fusion Res.* **6** 658

[19] Perogaro F, Bulanov S V, Califano F, Zh. Esirkepov T, Lontano M, Meyer-ter-Vehn J *et al* 1997 *Plasma Phys. Control. Fusion* **38** B26

[20] Singh M, Gopal K and Gupta D 2016 *Phys. Lett.* A **380** 1437

[21] Liseykina T V, Popruzhenko S V and Macchi A 2016 *New J. Phys.* **18** e072001

[22] Santos J J, Bailly-Crandvaux M, Ehret M, Arefiev A V, Batani D, Beg F N *et al* 2018 *Phys. Plasmas* **25** e056705

[23] Zeldovich Y B 1967 *Sov. Phys. JETP* **24** 1006

[24] Ritus V I 1967 *Sov. Phys. JETP* **24** 1041

[25] Oks E and Sholin G V 1977 *Opt. Spectrosc.* **42** 434

[26] Zhuzhunashvili A I and Oks E 1977 *Sov. Phys. JETP* **46** 1122

[27] Oks E, Dalimier E, Angelo P and Pikuz T 2024 *Open Physics* **22** 20240002

Chapter 5

Polarization and directional effects in the non-resonant coupling of the monochromatic and quasistatic electric fields in plasmas

In the electric field

$$\boldsymbol{E}(t) = \boldsymbol{E}_0 \cos \omega t + \boldsymbol{F}, \tag{5.1}$$

quasienergies of a hydrogenic atom or ion are determined by the eigenvalues of the following stationary matrix (see section 2.3 of book [1]):

$$\langle\langle \mathrm{O}n\alpha'|V|On\alpha\rangle\rangle = \langle n'_1 n'_2 m'|xF_x J_0(3nE_0/(2Z\omega)) + zF_z|n_1 n_2 m\rangle \tag{5.2}$$

Periodic parts of the so-called correct states of the zeroth order can be expressed as linear combinations of the following states:

$$|On\alpha\rangle\rangle = |n_1 n_2 m\rangle \exp[-(2Z\omega)^1{}^-3in(n_1 - n_2)E_0 \sin \omega t]. \tag{5.3}$$

The coefficients of the linear combinations of the states from equation (5.3) are found from the eigenvectors of the matrix (5.2).

By utilizing the ideas from work [3], it is possible to present analytically the quasienergy states for any principal quantum number n. This is because the same matrix as in equation (5.2) corresponds to the Stark effect in the following effective static field

$$\boldsymbol{F}_{\text{eff}}^{(n)} = (F_x J_0(3nE_0/2Z\omega),\ 0,\ F_z) \tag{5.4}$$

in the basis of the parabolic quantum numbers $|n_1 n_2 m\rangle$ with the z-axis chosen parallel to $\mathbf{E}_0$. For the static effective electric field $\boldsymbol{F}_{\text{eff}}^{(n)}$, the eigenvectors are the states of the

doi:10.1088/978-0-7503-6285-6ch5

parabolic $|n'_1 n'_2 m'\rangle$, where the new quantization axis z' is chosen parallel to $\boldsymbol{F}^{(n)}_{\text{eff}}$. The corresponding eigenvalues are (relative to ε_n):

$$\tilde{\varepsilon}_{n'_1 n'_2 m'} = 3n(n'_1 - n'_2)\boldsymbol{F}^{(n)}_{\text{eff}}/2Z. \tag{5.5}$$

Thus, the problem got significantly simplified. It is reduced to calculating the matrix transforming the parabolic wave functions while the quantization axis is rotated from the direction of $\mathbf{E}_0$ to the direction of $\boldsymbol{F}^{(n)}_{\text{eff}}$. The corresponding polarized profiles of the Lyman-alpha line in the z- and x-polarizations are as follows:

$$\begin{aligned}
I^{(e_z)}(\Delta\omega) = \sum_{k=-\infty}^{+\infty} \Big\{ & \delta(\Delta\omega - 2k\omega)2(\sin^2\varphi_2)J^2_{2k}(3\beta) \\
& + [\delta(\Delta\omega - 2k\omega - 3F^{(2)}_{\text{eff}}/Z) + \delta(\Delta\omega - 2k\omega + 3F^{(2)}_{\text{eff}}/Z)] \\
& \times \mathbf{J}^2_{2k}(3\beta)\cos^2\varphi_2 + [\delta(\Delta\omega - (2k-1)\omega - 3F^{(2)}_{\text{eff}}/Z) \\
& + \delta(\Delta\omega - (2k-1)\omega + 3F^{(2)}_{\text{eff}}/Z)]\mathbf{J}^2_{2k-1}(3\beta)\cos^2\varphi_2 \Big\}, \\
I^{(e_x)}(\Delta\omega) = \delta(\Delta\omega) & (\cos^2\varphi_2)/2 \\
& + [\delta(\Delta\omega - 3F^{(2)}_{\text{eff}}/Z) + \delta(\Delta\omega - 3F^{(2)}_{\text{eff}}/Z)]\sin^2\varphi_2; \\
\beta \equiv E_0/(\omega Z),\ & \varphi_n \equiv \arccos(F_z/F^{(n)}_{\text{eff}}).
\end{aligned} \tag{5.6}$$

In paper [3], the above results were applied to the analysis of the deuterium spectral lines emitted by the edge plasma of the tokamak T-10. The direction of the observation was parallel to the central chord in the equatorial plane of T-10. A polarizer was placed in the optical system.

Figure 5.1 presents the polarized experimental profiles of the deuterium-alpha line (curve 1 and 2) and the red half of the profile of the deuterium-gamma line (curve 3), obtained at the magnetic field $B_0 = 1.65$T, the electron density $N_e = 2.5 \times 10^{13}\text{cm}^{-3}$, and discharge current $J = 108$ kA. Curve 1 was recorded in the polarization perpendicular to $\mathbf{B}_0$, while curves 2 and 3 were recorded in the polarization parallel to $\mathbf{B}_0$.

The dips in the central part of the deuterium-alpha and deuterium-gamma lines were recorded in the profiles polarized parallel to $\mathbf{B}_0$—the π profiles. (Here the term 'dips' has nothing to do with the Langmuir-wave-caused dip structures discussed in chapter 4 of the present book.) These dips cannot be explicated by the relatively small Zeeman splitting or by a self-absorption of the emission (the latter is because there are no such dips in the deuterium-alpha and deuterium-gamma lines in the other polarization or in any of the deuterium-beta profiles).

It turned out that the properties of these polarization profiles can be explicated as being caused by the superposition of a high-frequency electric field $\boldsymbol{E}_0 \cos \omega t$, polarized along the observation direction, and a quasistatic electric field polarized parallel to $\mathbf{B}_0$:

$$\boldsymbol{E}\,(t) = \boldsymbol{e}_z E_0 \cos \omega t + \boldsymbol{e}_x F. \tag{5.7}$$

In equation (5.7), $\boldsymbol{e}_x$ and $\boldsymbol{e}_z$ are unit vectors.

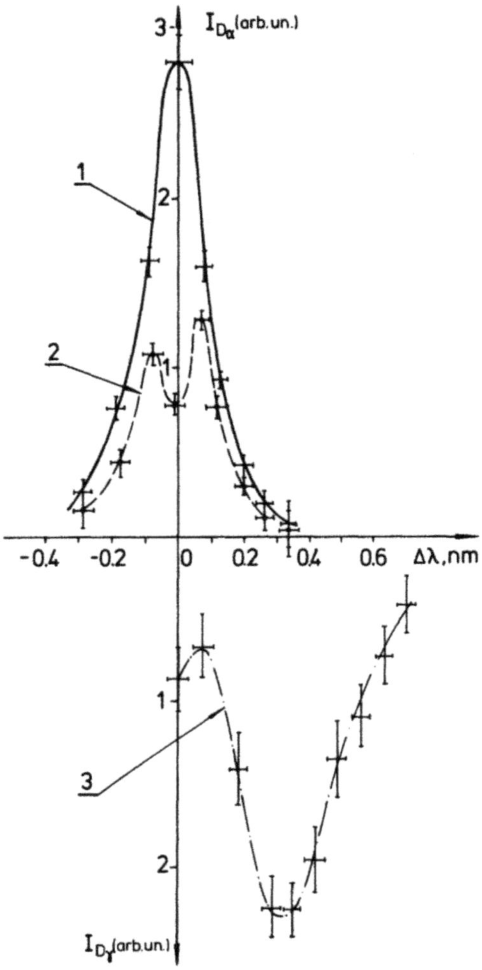

Figure 5.1. Polarized experimental profiles of the deuterium-alpha line (*curves 1 and 2*) and the red half of the profile of the deuterium-gamma line (*curve 3*), obtained at the magnetic field $B_0 = 1.65$T, the electron density $N_e = 2.5 \times 10^{13}$cm^{-3}, and discharge current $J = 108$ kA. Curve 1 was recorded in the polarization perpendicular to $\mathbf{B}_0$, while curves 2 and 3 were recorded in the polarization parallel to $\mathbf{B}_0$. Reprinted from [1], copyright (1995), with permission from Springer Nature.

The fact that dips showed up in the π-profiles of the deuterium- alpha and deuterium-gamma lines but there was no dip in the π-profile of the deuterium-beta line can therefore be explicated by the partial suppression by the field $\boldsymbol{E}_0 \cos \omega t$ of the Stark splitting of the levels of $n = 2$, 3, 5, and the *complete* suppression of the $n = 4$ level: $F_{\text{eff}}^{(4)} = 0$. Indeed, if $F_{\text{eff}}^{(4)} = 0$, then the most intense π-components of the quasistatic Stark profile of the deuterium-beta line are radiated at the unshifted frequency, that is, at $\Delta\omega = 0$, which is why there is no central dip in this spectral line.

By assuming that the argument of the Bessel function in $F_{\text{eff}}^{(4)} = 0$ is equal to the first zero of the Bessel function, it was deduced $E_0/(m_e e\omega) \approx 0.40$. Further, since

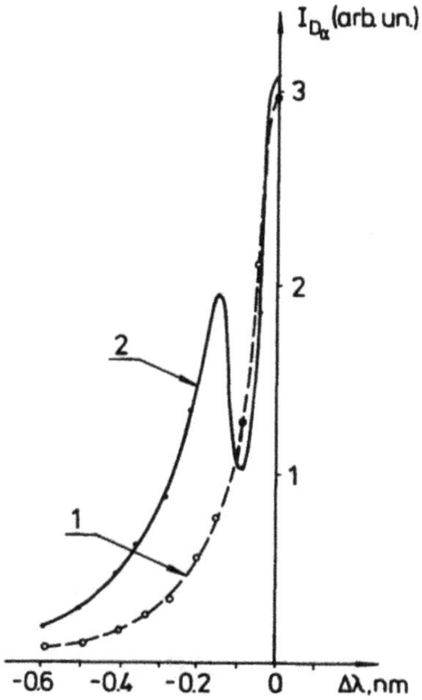

Figure 5.2. Profiles of the blue half of the deuterium-alpha line recorded at $J = 450\text{kA}$, $B_0 = 3.05\text{T}$, and $N_e = 6 \times 10^{13}\text{cm}^{-3}$. Curve 1: the polarization parallel to $\mathbf{B}_0$; curve 2—the unpolarized profile. Reprinted from [1], copyright (1995), with permission from Springer Nature.

the frequency ω was supposed to be near the electron cyclotron frequency ω_{Be} at the magnetic field $B_0 = 1.65\text{T}$, it was determined that $E_0 \approx 14$ kV cm^{-1}. Finally, from the π-profile of the deuterium-alpha line it was deduced that the quasistatic field was $F \approx 20$ kVcm^{-1}. Since the σ-profiles of the deuterium-alpha and deuterium-beta lines did not show intense satellites at the frequencies $\Delta\omega = \pm\omega, \pm 2\omega$, then the direction of the field $\boldsymbol{E}_0 \cos \omega t$ should have been approximately along the direction of the observation.

At the higher magnetic field $B_0 = 3.05\text{T}$ there occurred a very significant modification of the profiles of the deuterium lines. Namely, the σ-profile of the deuterium-alpha line, rather than the π-profile, demonstrated the larger broadening and the central dip—see figure 5.2. This meant that the quasistatic field was directed along the y-axis, that is, perpendicular to both the direction of observation and to $\boldsymbol{B}_0$. The quantitative analysis leads to the following results: $F \approx 20$ kVcm^{-1} and $E_0 < 10$ kV cm^{-1}.

References

[1] Oks E 1995 *Plasma Spectroscopy: The Influence of Microwave and Laser Fields* (Berlin: Springer)

[2] Gavrilenko V P and Oks E 1985 *Proc. 17th Int. Conf. on Phenomena in Ionized Gases (Budapest)* p 1081

[3] Gavrilenko V P, Oks E and Rantsev-Kartinov V A 1986 *JETP Lett.* **44** 404

Chapter 6

Polarization and directional effects in the anisotropic Stark broadening of hydrogen spectral lines due to the motion of the radiating ions

In this chapter we analyze anisotropic effects in the dynamical Stark broadening by ions in plasmas. Seidel [1] was the first to consider the situation where for a radiating atom moving with a constant velocity, in the reference frame of the radiating atom, the plasma appears *anisotropic*. Consequently, the spectral line profiles depend both on the velocity of the radiating atom and on the direction of observation. The corresponding changes in the spectral line profiles are especially significant in the case where the dynamical Stark broadening is caused by plasma ions that are much heavier than the radiating atom [1]. The analytical calculations in paper [1] were performed in frames of the Conventional Theory (CT) of the Stark broadening [2] and therefore had the divergency at small impact parameters.

In paper [3] the authors studied analytically the same problem by using the advanced formalism [4, 5] where there is no divergency at small impact parameters. By using the hydrogen Lyman-alpha line as an example, the authors of paper [3] demonstrated that the anisotropic character of the Stark broadening led to the *narrowing* of the line profile. Below are some details.

If the perturbing ions in the plasma are much heavier than the radiating atom, then the velocity of the perturbing ions can be disregarded compared to the velocity **v** of the radiating atom. In this situation, in the reference frame of the radiating atom, the perturbing ions are seen as the *ion beam* having the velocity—**v**. Of course, for different radiating atoms, the perceived ion beams have different velocities—both by the magnitude and by the direction.

doi:10.1088/978-0-7503-6285-6ch6

In paper [3] the calculations were performed for the limiting case of the ion impact broadening—for revealing the anisotropic effects in the simplest way. Situations where the perturbing ions produce the impact broadening are well-described in the literature—for example, in works [6–9]. It is worth noting that in the situation where both the plasma ions and the plasma electrons produce the impact broadening, the contribution of the plasma electrons to the Stark width can be disregarded compared to the contribution of the plasma ions to the Stark width because the latter contribution is by a factor of v_{Te}/v_{Ti} greater than the former (except if $T_e \gg T_i$).

With the z-axis along the direction of the ion beam, the Hamiltonian can be expressed as follows

$$H(t) = H_1(t) + V(t), \;\; H_1(t) \equiv H_0 - d_z E_z(t), \;\; V(t) \equiv -dE_x - d_y E_y. \tag{6.1}$$

In equation (6.1), $E(t)$ is the ion electric field. The part $H_1(t)$ of the total Hamiltonian was allowed for exactly by diagonalizing in the parabolic quantization. The rest of the total Hamiltonian, that is $V(t)$, was processed by means of the Dyson perturbation expansion.

The initial formula for the profile of the spectral line was

$$I(\omega, v) = -\frac{1}{\pi} \operatorname{Re} \sum_{\sigma} \sum_{\alpha\alpha'\beta\beta'} \langle\beta|d_\sigma|\alpha\rangle\langle\alpha'|d_\sigma|\beta'\rangle \, \langle\langle\alpha\beta|G^{-1}|\alpha'\beta'\rangle\rangle. \tag{6.2}$$

The spectral line profile in equation (6.2) depends on the ion beam velocity. This is because the spectral operator G depends on the velocity:

$$G = i\Delta\omega + \Phi_{ab}(v), \tag{6.3}$$

where $\Phi_{ab}(v)$ is the velocity-dependent impact operator and $\Delta\omega$ is the detuning from the unperturbed frequency of the spectral line.

For the anisotropic Stark broadening, the impact operator has a relatively simple form:

$$\Phi_{ab} = N_i v \int_0^\infty 2\pi\rho \, d\rho \left\{ S_a S_b^* - 1 \right\}_{\vec{\rho}}. \tag{6.4}$$

It should be underscored again that in the reference frame of the radiating atom, all heavy perturbing ions have the same velocity, which is why in equation (6.4) there is no average over the velocities.

In distinction to the CT, in paper [3] the scattering matrix S was expressed as follows:

$$\begin{gathered} S = \exp\left[(i/\hbar)\int_\infty^\infty dt \; d_z E_z(t)\right] \hat{T} \exp\left[(i/\hbar)\int_\infty^\infty dt \; Q^*(d_x E_x + d_y E_y) Q\right], \\ Q = \exp\left[-(i/\hbar)(H_0 t - \int_\infty^t dt' \; d_z E_z(t'))\right]. \end{gathered} \tag{6.5}$$

The physical meaning of equation (6.5) is that the projection of the field of perturbing ions on the relative perturber-radiator velocity was taken into account exactly.

For the Lyman lines, $S_b = 1$, so that the impact operator from equation (6.4) gets simpler. In paper [3] for the matrix elements of the impact operator for the Lyman lines, the following expression was obtained by means of the modified Dyson expansion in its second order:

$$\Phi_{\alpha\alpha'} = -4\pi N \frac{Z_i^2 e^2}{\hbar^2 v_0} \frac{1}{u} \sum_{\alpha''} d^x_{\alpha\alpha''} d^x_{\alpha''\alpha'} \int_0^\infty C_\pm(Z) \frac{dZ}{Z}. \tag{6.6}$$

In equation (6.6), Z_i is the charge of perturbing ions, $v_0 = (2T_a/m_p)^2$ is the mean thermal velocity of the radiating atoms, $u = v/v_0$ is a reduced velocity, T_a is the atom temperature, and m_p is the mass of the proton. The integration over Z means to the averaging over impact parameters. The impact operator in equation (6.6) has only the nonadiabatic terms—in distinction to the isotropic Stark broadening.

In equation (6.6), C_- and C_+ are the broadening functions for the diagonal and nondiagonal matrix elements, respectively. They were expressed as follows:

$$C_\pm(Z) = \frac{1}{2} \int_\infty^\infty \int_\infty^{x_1} \frac{dx_1 dx_2}{[g(x_1)g(x_2)]^3} \exp\left[\frac{i}{Z}(1/g(x_1) \pm 1/g(x_2))\right], \qquad g(x) \equiv \sqrt{1+x^2}. \tag{6.7}$$

For the real parts of these broadening functions, the double integrals in equation (6.7) were calculated in paper [3] analytically:

$$Re\, C_\pm(Z) = \left[1 - \frac{\pi}{2} H_1(1/Z)\right]^2 \mp \left[\frac{\pi}{2} J_1(1/Z)\right]^2. \tag{6.8}$$

In equation (6.8), $J_1(1/Z)$ and $\boldsymbol{H}_1(1/Z)$ are the Bessel and Struve functions, respectively.

The asymptotics of $ReC_\pm$ at large and small values of Z are as follows:

$$ReC_\pm \approx 1, \ Z \gg 1, \tag{6.9}$$

$$ReC_+ \approx (\pi Z/2)\sin(1/Z), \ Z \ll 1, \tag{6.10}$$

$$ReC_- \approx \pi Z/2, \ Z \ll 1. \tag{6.11}$$

From equations (6.10) and (6.11), it is easy to see that the integration over Z in equation (6.6) indeed converges at small impact parameters.

The divergency at large Z, just as in the CT, can be dealt with by engaging the cutoff at the Debye radius:

$$Z_u \equiv Z_0 u, \ Z_0 \equiv \frac{\hbar v_0 \rho_{De}}{Z_i e \,|\,(d_z)_{\alpha\alpha} - (d_z)_{\alpha'\alpha'}\,|}. \tag{6.12}$$

In paper [3] there was explicitly calculated the corresponding profile of the hydrogen Lyman-alpha line:

$$I_u(\omega,\ v) = \frac{1}{3\pi}\left(\frac{\Gamma_\pi}{\Delta\omega^2 + \Gamma_\pi^2} + \frac{2\,\Gamma_\sigma}{\Delta\omega^2 + \Gamma_\sigma^2}\right), \tag{6.13}$$

In equation (6.13),

$$\Gamma_\sigma = (\eta_0/u)\mathrm{Re}\int_0^{Z_0 u} C_-(Z)dZ/Z, \Gamma_\pi = (\eta_0/u)\mathrm{Re}\int_0^{Z_0 u}(C_-(Z) - C_+(Z))dZ/Z. \tag{6.14}$$

In equation (6.14),

$$\eta_0 = 4\pi N_i Z_i^2 \hbar^2/\left(m_e^2 v_0\right) \approx 1.21 \times 10^{-5} Z_i^2\left[N_i(cm^{-3})\right]\left[T_a(eV)\right]^{-2} \tag{6.15}$$

A remarkable property of the π-component of the Lyman-alpha line is that it does not need not only the lower cutoff, but also the upper cutoff for the integration over the impact parameters. This is because the impact Stark width of the π-component of the Lyman-alpha line is controlled by the difference between the nondiagonal diagonal matrix elements of the impact operator. The resulting simple expression as follows:

$$\Gamma_\pi = \frac{\pi^2}{4}\eta_0\frac{1}{u}. \tag{6.16}$$

Equations (6.13)–(6.15) display the velocity-dependent profile of the spectral line emitted by a single radiating atom. For obtaining the profile of the spectral line emitted by the entire ensemble of radiating atoms, the spectral line profile from equation (6.13) has to be averaged over atomic velocities. Here it is important to emphasize that this is the primary distinction from the isotropic situation where to the averaging over velocities is subjected the impact broadening operator rather than the spectral line profile.

In paper [3] for showing the results of the anisotropic impact broadening in the purest form, the authors disregarded the Doppler effect and performed the velocity averaging as follows:

$$I(\omega) = \int_0^\infty I_u(\omega,\ v)W_M(\vec{v})d\vec{v}, \tag{6.17}$$

where $W_M(v)dv = 4\pi^{-1/2}u^2\exp(-u^2)du$. The corresponding result is not just a theoretical exercise: it is relevant for experiments employing the Doppler-free two-photon fluorescence. It is also relevant to those laser fusion plasmas where the Stark broadening dominates over the Doppler broadening.

For the π-component of the Lyman line, in paper [3] there was explicitly performed the integration in equation (6.17) and they obtained the following:

$$I_\pi(\Delta\omega) = \frac{8}{\pi^{7/2}\eta_0}\frac{1}{\mu^2}G\left(1,0,\frac{1}{\mu^2}\right),\ \mu \equiv \Delta\omega/\left(\frac{\pi^2}{4}\eta_0\right), \tag{6.18}$$

where $G(a,b;x)$ is the confluent hypergeometric function of the second kind.

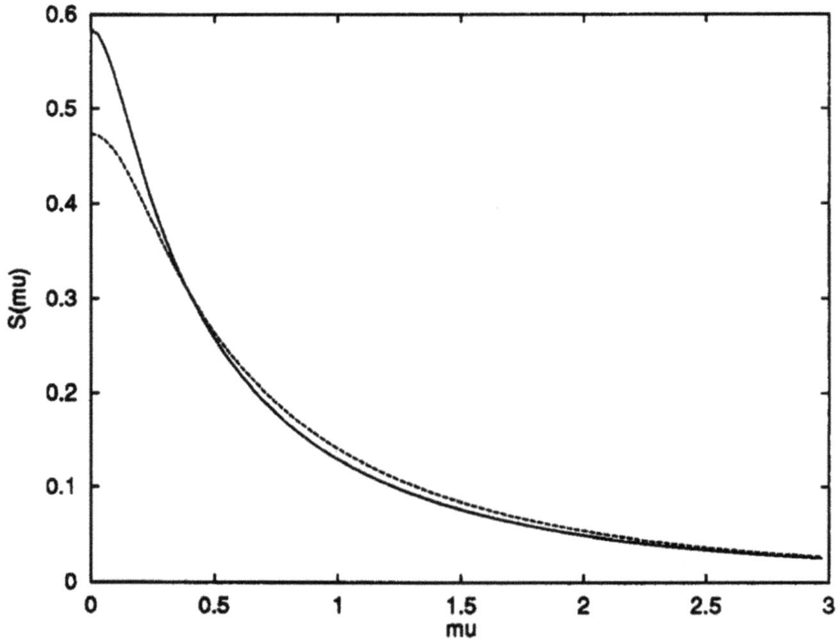

Figure 6.1. Scaled profiles of the hydrogen Lyman-alpha line calculated within the advanced formalism, employed in paper [3], in two different ways: (1) with the velocity averaging applied to the impact operator from equation (6.8), (the 'isotropic' approach)—dashed line; (2) with the velocity averaging applied to the spectral line profile from equations (6.13) and (6.15), (the 'anisotropic' approach)—solid line. The profiles are normalized to unity; the abscissa is measured in units of $\mu = \Delta\omega/[\pi^2\eta_0/4]$, where η_0 was defined in equation (6.15). Reprinted from [3], copyright (1995), with permission from Elsevier.

For pictorial examples, the authors of paper [3] calculated the profiles of the hydrogen Lyman-alpha line for plasma parameters $N_e = 10^{13}$ cm^{-3}, $T = 100$ eV, $Z_i = 1$. These plasma parameters are relevant for the edge plasmas of tokamaks and stellarators.

Figure 6.1 exhibits the quantitative importance of the anisotropic Stark broadening by itself. Namely, it shows the profiles of the Lyman-alpha line calculated in frames of the same advanced formalism (employed in paper [3]) in two ways: (1) with the velocity averaging applied to the impact operator from equation (6.8), (the 'isotropic' approach); (2) with the velocity averaging applied to the spectral line profile from equations (6.13) and (6.15), (the anisotropic' approach).

The comparison of the two calculated profiles demonstrates that the allowance for the anisotropic character of broadening just by itself results in the Stark width 30% smaller compared to the Stark width in the isotropic case.

Then, for exhibiting the quantitative importance of utilizing the advanced formalism, rather than the CT, the authors of paper [3] displayed the hydrogen Lyman-alpha profiles calculated in the anisotropic approach in two different ways—see figure 6.2: (1) by employing the advanced formalism (solid line); (2) by utilizing the CT (dashed line).

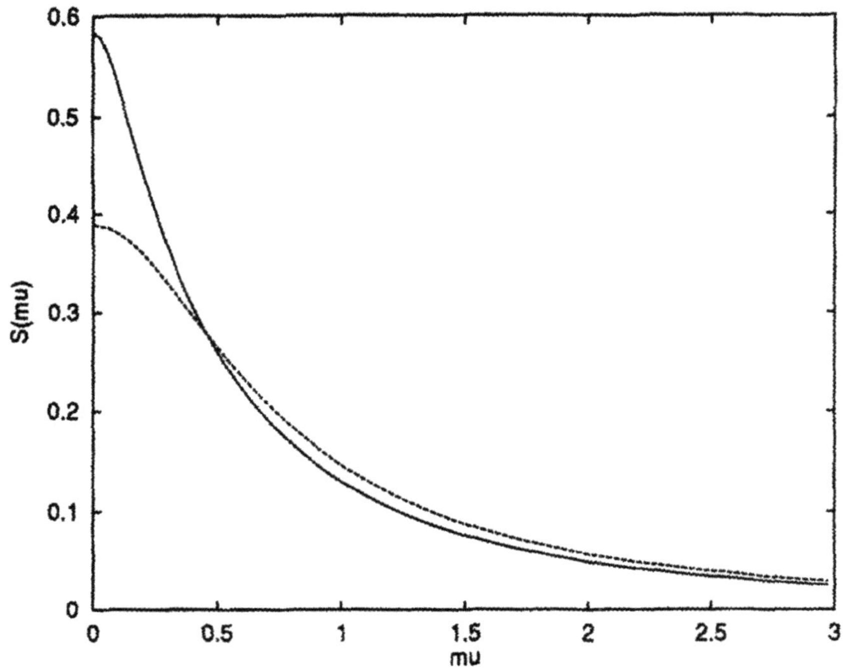

Figure 6.2. Profiles of the hydrogen Lyman-alpha calculated in the anisotropic approach in two different ways: (1) by employing the advanced formalism (solid line); (2) by utilizing the CT (dashed line). The profiles are normalized to unity and the abscissa is measured in the same units as in figure 6.1. Reprinted from [3], copyright (1995), with permission from Elsevier.

The comparison of the two calculated profiles in figure 6.2 demonstrates the following. The usage of the CT leads to the error of 60% in the Stark width: the profile calculated by employing the advanced formalism is very significantly narrower than the profile calculated by using the CT.

References

[1] Seidel J 1979 *Z. Naturforsch.* **34a** 1389

[2] Griem H R 1974 *Spectral Line Broadening by Plasmas* (New York: Academic)

[3] Derevianko A and Oks E 1995 *J. Quant. Spectrosc. Radiat. Transfer* **54** 137

[4] Ispolatov Y and Oks E 1994 *J. Quant. Spectrosc. Radiat. Transfer* **51** 129

[5] Oks E 2006 *Stark Broadening of Hydrogen and Hydrogenlike Spectral Lines in Plasmas: The Physical Insight* (Oxford: Alpha Science International)

[6] Derevianko A and Oks E 1994 *Phys. Rev. Lett.* **73** 2059

[7] Lisitsa V S 1977 *Sov. Phys.-Usp.* **122** 449

[8] Stehle C and Feautrier N 1984 *J. Phys.* B **17** 1477

[9] Gaisinsky I M and Oks E A 1985 *J. Phys.* B **18** 1449

Chapter 7

Polarization and directional effects in the radiation of plasma-based x-ray lasers

One avenue in designing soft x-ray lasers is grounded on the recombination pumping of hydrogen-like ions—the ions from which all electrons, except one, were removed by the optical field ionization [1–3]. The gain of x-ray lasers is controlled by the product of the oscillator strength for the lasing transition and the halfwidth of the lasing line. In papers [4, 5] there was demonstrated the possibility to significantly diminish the Stark width of the hydrogen-like spectral lines by the application of a high frequency electric field: the latter modifies the coupling of the electric microfield of a plasma with the hydrogen-like ion. In particular, in paper [6] it was revealed that by applying a linearly-polarized field of an optical laser, it is possible—for some x-ray lasing hydrogenic spectral lines—to substantially narrow the profile of the absorption coefficient and in this way to increase the gain of the x-ray laser. Later, this narrowing effect was also entertained in paper [7].

Then in paper [8] it was considered how the elliptically-polarized intense field of an optical laser (EPIFOL) affects the Stark broadening of some x-ray spectral lines of hydrogen-like ions. The analysis in paper [8] was based on the analytical results from paper [9]. The model microfield method was utilized for describing the interaction of the plasma electric microfields with a hydrogen-like atom. It was demonstrated that the employment of the EPIFOL has the following two advantages. First, the gain of some x-ray lasing transitions will be substantially increased. Second, it would open up the possibility to create a tunable x-ray laser—the laser tunable in a broad range of frequencies. Below are some details.

A hydrogen-like ion of a nuclear charge Z is considered under the joint action of the plasma microfield $\overrightarrow{F}(t)$and the EPIFOL of the form

$$\overrightarrow{E}(t) = \varepsilon_0[\overrightarrow{e_z} \cos(\omega_0 t + \gamma) + \xi\, \overrightarrow{e_x} \sin(\omega_0 t + \gamma)], \tag{7.1}$$

doi:10.1088/978-0-7503-6285-6ch7

where ξ is the ellipticity degree; $\vec{e_x}$ and $\vec{e_z}$ are the unit vectors. For the principal quantum number $n = 2$, the corresponding wave functions have been obtained in paper [9]:

$$\Psi_1 \equiv \tilde{\psi}_1(t), \quad \Psi_2 \equiv \tilde{\psi}_2(t), \quad \Psi_3 \equiv \exp(-i\kappa t)\tilde{\psi}_3(t), \quad \Psi_4 \equiv \exp(i\kappa t)\tilde{\psi}_4(t),$$

$$\tilde{\psi}_k(t) = 2^{-1}\{(-1)^{k+1}(\phi_1 - \phi_2) + i\phi_3 \exp[-i\beta(t)] + i\phi_4 \exp[i\beta(t)]\},$$

$$\tilde{\psi}_p(t) = 2^{-1}\{i(\phi_1 + \phi_2) + (-1)^{p+1}\{\phi_3 \exp[-i\beta(t)] - \phi_4 \exp[i\beta(t)]\}\}, \tag{7.2}$$

$$k = 1, 2; \quad p = 3, 4,$$

where

$$\kappa = \omega_0 \xi w J_1(w), \quad \beta = w \sin(\omega_0 t + \gamma), \quad w = 3\varepsilon_0/(Z\omega_0), \tag{7.3}$$

In equation (7.2), the functions ϕ_s (s= 1, 2, 3, 4) are the following parabolic wave functions $\phi_1 \equiv |001\rangle, \phi_2 \equiv |00-1\rangle, \phi_3 \equiv |100\rangle, \phi_4 \equiv |010\rangle$, labeled by the parabolic quantum numbers n_1, n_2, m. The quantization axis is Oz. In equation (7.3), $J_1(w)$ is the Bessel function. The above expressions for these wave functions are valid for $|\xi| \leqslant w^{-1/2}$. Here and below, atomic units $\hbar = m_e = e = 1$ are utilized.

The corresponding Schrödinger equation is:

$$i\frac{\partial \Phi}{\partial t} = [H_a + z\varepsilon_0 \cos(\omega_0 t + \gamma) + x\varepsilon_0 \xi \sin(\omega_0 t + \gamma) + V(t)]\Phi, \quad V(t) \equiv \vec{r} \cdot \vec{F}(t), \tag{7.4}$$

In equation (7.4). H_a is the unperturbed Hamiltonian.

The solution was sought in the form (for $n = 2$)

$$\Phi_j(t) = \sum_r T_{rj}(t, 0)\tilde{\psi}_r(t). \tag{7.5}$$

In equation (7.5), $T_{rj}(t, 0)$ is the time evolution operator. The initial conditions have been $T_{rj}(0, 0) = \delta_{rj}$ (δ_{rj} being the Kronecker symbol).

On substituting equation (7.5) in equation (7.4), the authors of paper [9] got

$$\frac{dT}{dt} = M(\vec{F}, t)T. \tag{7.6}$$

In equation (7.6), the matrix elements of the operator $M(\vec{F}, t)$ were

$$M_{qr}(\vec{F}, t) = -i\langle\tilde{\psi}_q(t)|V(t)|\tilde{\psi}_r(t)\rangle. \tag{7.7}$$

For utilizing the Model Microfield Method (MMM) for expressing the effect of the plasma microfield, it was necessary to find the evolution operator for the situation where the field $\vec{F}$ was static: the static evolution operator T_S. The authors of paper [9] dealt with the case where the Stark splitting of the of the level $n = 2$ is much smaller than the frequency ω_0 of the optical laser:

$$\omega_0 \gg 3F/Z. \tag{7.8}$$

In this situation, the authors of paper [9]—for solving equation (7.6) for the time-independent field $\overrightarrow{F}$—employed the averaging method by Krylov–Bogoliubov–Mitropolskii [10, 11]: the matrix elements $M_{qr}(\overrightarrow{F}, t)$ are substituted by their time-averaged values $\overline{M}_{qr}(\overrightarrow{F})$:

$$M_{qr}(\overrightarrow{F}, t) \rightarrow \overline{M}_{qr}(\overrightarrow{F}), \quad \overline{M}_{qr}(\overrightarrow{F}) = \frac{\omega_0}{2\pi} \int_0^{2\pi/\omega_0} dt \; M_{qr}(\overrightarrow{F}, t), \tag{7.9}$$

resulting in the equation

$$\frac{dT_S}{dt} = \overline{M}(\overrightarrow{F})T_S, \tag{7.10}$$

where

$$\overline{M}(\overrightarrow{F}) = \begin{pmatrix} if_y & 0 & if_x - f_z & if_x + f_z \\ 0 & -if_y & if_x - f_z & if_x + f_z \\ if_x + f_z & if_x + f_z & -i\kappa & 0 \\ if_x - f_z & if_x - f_z & 0 & i\kappa \end{pmatrix}. \tag{7.11}$$

In equation (7.11),

$$f_x = \frac{3}{2Z} J_0(w)F_x, \quad f_y = \frac{3}{Z} J_0(w)F_y, \quad f_z = \frac{3}{2Z} F_z, \tag{7.12}$$

where $J_0(w)$ is the Bessel function.

In the model microfield method [12], the key role is played by the jumping frequency $\nu(\overrightarrow{F})$ and the distribution function $P(\overrightarrow{F})$ [9]. Between two adjacent jumps, the field $\overrightarrow{F}$ is considered stationary.

The Fourier transform of the evolution operator averaged over the stochastic realizations of the plasma microfield $\overrightarrow{F}(t)$ was, according to paper [9], the following:

$$\tilde{T}_{\mathrm{MMM}}(\omega) = \{\tilde{T}_S(\tilde{\omega})\}_{av} + \{\nu\tilde{T}_S(\tilde{\omega})\}_{av}\{\nu I - \nu^2\tilde{T}_S(\tilde{\omega})\}_{av}^{-1}\{\nu\tilde{T}_S(\tilde{\omega})\}_{av}. \tag{7.13}$$

In equation (7.13), $\tilde{T}_S(\tilde{\omega})$ is the Laplace transform of the static operator $T_S(t)$ calculated at $\tilde{\omega} = \omega + i\nu(\overrightarrow{F})$ for a static field $\overrightarrow{F}$; the symbol $\{\ldots\}_{av}$ stands for the average over the probability distribution $P(\overrightarrow{F})$, and $\nu = \nu(\overrightarrow{F})$.

Then, the spectral line profile of the Lyman-alpha line can be written as follows:

$$I^{(\zeta)}(\omega) = \frac{1}{\pi} \sum_{\alpha',\alpha''} \mathrm{Re}\,[\widetilde{T}_{\mathrm{MMM}}(\omega)]_{\alpha'',\alpha'} R^{(\zeta)}_{\alpha'\alpha''}. \tag{7.14}$$

In equation (7.14), ζ labels the polarization of the emitted light ($\zeta = x, y, z$), and

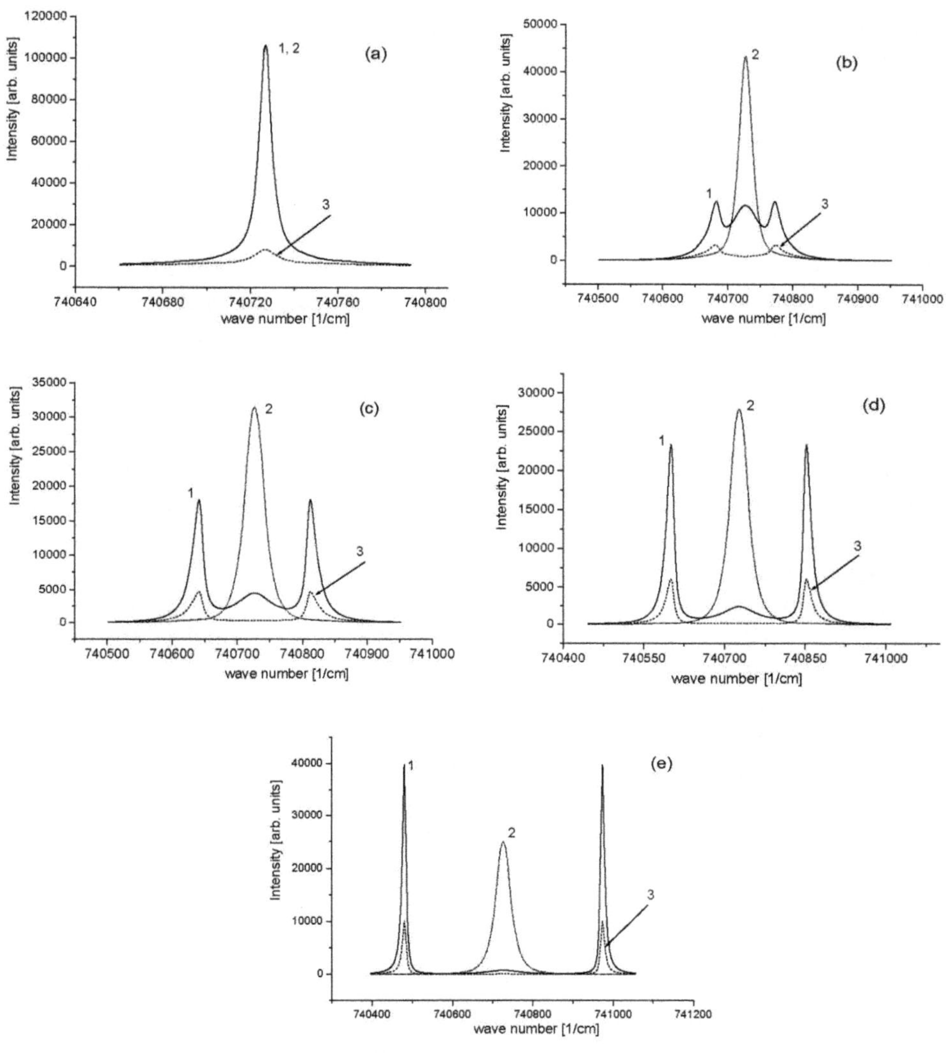

Figure 7.1. Calculated profiles of the Li III Lyman-alpha line (the wavelength 13.5 nm) under the elliptically-polarized electric field of the CO_2 laser in a plasma of the temperature $T_e = T_i = 3$ eV and of the electron density $N_e = 5.0 \times 10^{18}$ cm^{-3} (the plasma parameters corresponding to the experiment [2]). The laser amplitude is 3.6 MV cm^{-1}. The plasma ions are Li^{3+}. The subcases a, b, c, d, and e are for the ellipticity degree 0, 0.05, 0.10, 0.15, and 0.20, respectively. Profiles 1, 2, and 3 in each subcase indicate the x-ray radiation polarized in *x*-, *y*-, and *z*-directions, respectively, *z* being the direction of the major axis of the ellipse, while *x* being the direction of the minor axis of the ellipse. Reprinted from [8], copyright (2004), with permission from Springer Nature.

$$R^{(\zeta)}_{\alpha\,\alpha''} = \left\{ \left\langle \phi_0 | \zeta | \tilde{\psi}^{(0)}_{\alpha''}(\tau) \right\rangle \left\langle \tilde{\psi}^{(0)}_{\alpha'}(0) | \zeta | \phi_0 \right\rangle \right\}_{\gamma}. \qquad (7.15)$$

In equation (7.15), ϕ_0 is the wave function of the level $n = 1$, $\tilde{\psi}^{(0)}_{\alpha}(\tau)$ is the zero harmonic of the periodic wave function $\tilde{\psi}_{\alpha}(\tau)$, and $\{\dots\}_{\gamma}$stands for the average over the initial phase γ of the laser field $\overrightarrow{E}(t)$.

As an example, the authors of paper [8] calculated the profiles of the Li III Lyman-alpha line (the wavelength 13.5 nm) under the elliptically-polarized electric field of the CO_2 laser in a plasma of the temperature $T_e = T_i = 3$ eV and of the electron density $N_e = 5.0 \times 10^{18}$ cm^{-3} (the plasma parameters corresponding to the experiment [2]). They used formulas from paper [13] for calculating the distribution function $P(\overrightarrow{F})$—the formulas employing the results from papers [14–16].

Figure 7.1 demonstrates the modification of the profiles of the Lyman-alpha line of Li III in x-, y-, and z-polarizations, as the adiating ion is under the field of the CO_2 laser—for several values of the ellipticity degree in the range from $\xi = 0$ to $\xi = 0.2$. For the plasma parameters under the consideration, the dominating contribution to the Stark broadening of this spectral line is from the plasma ion microfield.

The profile polarized in the direction of the minor component of the dressing field (the x-polarization) is split in the following three components: the central (unshifted) component at the unperturbed position $\omega_{21}^{(0)}$ of the spectral line and two side components at the frequencies $\omega_{21}^{(0)} \pm \kappa$, κ being defined in equation (7.3). At the values of the ellipticity degree greater or equal to 0.05, the maxima of the lateral components exceed the maximum of the central component. Thus, it is feasible generating the x-ray radiation, polarized in the x-direction, at the shifted frequencies $\omega_{21}^{(0)} \pm \kappa$. Due to the fact that the quasienergy κ is proportional to the ellipticity degree ξ (see equation (7.3)), it becomes possible to design a *tunable* x-ray laser, the tuning being achieved by varying the ellipticity degree ξ.

So, under the EPIFOL, the x-ray radiation generated at the frequencies $\omega_{21}^{(0)} \pm \kappa$ will be polarized along the minor axis of the ellipse (in the x-direction). This is the distinction from the situation where the dressing field is linearly-polarized—then the polarization of the generated x-ray radiation is in the direction perpendicular to the dressing field.

References

[1] Nagata Y, Midorikawa K, Kubodera S, Obara M, Tashiro H and Toyoda K 1993 *Phys. Rev. Lett.* **71** 3774

[2] Donnelly T D, Da Silva L, Lee R W, Mrowka S, Hofer M and Falcone R W 1996 *J. Opt. Soc. Am.* B **13** 185

[3] Korobkin D V, Nam C H, Suckewer S and Goltsov A 1996 *Phys. Rev. Lett.* **77** 5206

[4] Gavrilenko V P and Oks E 1985 *Proc. 17th Int. Conf. on Phenomena in Ionized Gases* (Budapest: Hungarian Academic Science) p 1081

[5] Gavrilenko V P and Oks E 1989 *Proc. 19th Int. Conf. on Phenomena in Ionized Gases (Belgrade, Yugoslavia)* p 354

[6] Oks E 2000 *J. Phys. B: At. Mol. Opt. Phys.* **33** L801

[7] Alexiou S 2001 *J. Quant. Spectrosc. Radiat. Transfer* **71** 139

[8] Gavrilenko V P and Oks E 2004 *Eur. Phys. J.* D **28** 253

[9] Oks E and Gavrilenko V P 1983 *Opt. Commun.* **46** 205

[10] Krylov N N and Bogoliubov N N 1947 *Introduction to Non-linear Mechanics* (Princeton, NJ: Princeton University Press)

[11] Bogoliubov N N and Mitropolskii Y M 1961 *Asymptotic Methods in the Theory of Nonlinear Oscillations* (New York: Gordon and Breach)

[12] Brissaud A and Frisch U 1971 *J. Quant. Spectrosc. Radiat. Transfer* **11** 1767
[13] Golosnoy I O 1993 *Matematicheskoe Modelirovanie [Math. Modeling]* **5** 11 (in Russian)
[14] Iglesias C A and Lebowitz J L 1984 *Phys. Rev.* A **30** 2001
[15] Iglesias C A, DeWitt H E, Lebowitz J L, MacGowan D and Hubbard W B 1985 *Phys. Rev.* A **31** 1698
[16] Brissaud A, Goldbach C, Léorat J, Mazure A and Nollez G 1976 *J. Phys. B: At. Mol. Phys.* **9** 1129

IOP Publishing

Polarization and Directional Effects in the Radiation from Plasmas

Eugene Oks

Chapter 8

Polarization and directional effects in the Stark broadening of spectral lines by a relativistic electron beam in magnetic fusion plasmas

8.1 The case of the relativistic electron beam

In this section we analyze the Stark broadening of spectral lines by a Relativistic Electron Beam (REB) and polarization and directional effects that it causes. The interaction of an REB with plasmas is important—from the fundamental point of view—for understanding physics of plasmas. This study has also a practical significance for inertial fusion (also known as laser fusion), for heating of plasmas, for acceleration of charged particles in plasmas, as well as for the generation of intense coherent microwave radiation [1–3].

Among practical applications, the last but not least concerns magnetic fusion (specifically tokamaks) and has to do with runaway electrons. In some discharges due to a decay of plasma current, the latter is replaced electrons (called runaway electrons) that attain relativistic energies. This is dangerous especially for ITER—the next generation tokamak [4–6]. At various existing tokamaks, the measurements of the energy of the runaway electrons yielded ~0.2–10 MeV; the density of the runaway electrons was found experimentally to be in the ratio $\sim 10^{-1}$–10^{-4} to the bulk plasma electron density [7–9].

This is the reason for developing diagnostics of an REB and of the REB–plasma interaction. Specifically, for tokamaks, it is crucially important to detect the development of an REB in the timely fashion to mitigate the problem.

Here we follow paper [10] to present the analytical theory of the REB-caused Stark broadening of spectral lines of hydrogen or deuterium, as well as its application to the edge plasmas of tokamaks. An important fact is that under an REB, the dynamical Stark broadening of spectral lines in plasmas becomes

doi:10.1088/978-0-7503-6285-6ch8

anisotropic. Previous results on the cases of the anisotropic dynamical Stark broadening in plasmas are presented in chapter 6 of this book.

We note in passing that Rosato *et al* [11] tried to analyze the Stark broadening of the hydrogen Lyman-alpha line by an REB for the conditions of the edge plasmas of tokamaks. Unfortunately, the authors of paper [11] utilize the quasistatic approximation—despite it being absolutely inappropriate for the Stark broadening by fast electrons (such as the electrons of an REB); moreover, for the edge plasmas of tokamaks, the quasistatic approximation is inappropriate even for the Stark broadening by thermal electrons.

In paper [10] the authors employed the formalism from paper [12], but with two major differences from paper [12]. Namely, first, while in paper [12] there was considered the anisotropic dynamical Stark broadening by plasma ions, in paper [10] the authors analyzed the dynamical Stark broadening by the electrons of the REB. Second, in paper [1] the particles of the beam (the electrons) were considered to have relativistic energies.

In paper [10] the z-axis was chosen in the direction of the REB. Then the Hamiltonian $H(t)$ was presented in the form:

$$H(t) = H_1(t) + V(t),\ H_1(t) \equiv H_0 - d_z E_z(t),\ V(t) \equiv -dE_x - d_y E_y. \tag{8.1}$$

The part $H_1(t)$ was taken into account exactly by the diagonalization in the parabolic quantization. The rest of the interaction, that is $V(t)$, was allowed for by using the Dyson perturbation expansion.

The initial formula for the shape of the hydrogen or deuterium spectral line $I(\omega,v)$ has the dependence on the velocity v of the REB:

$$I(\omega, v) = -\frac{1}{\pi} Re \sum_{\sigma} \sum_{\alpha\alpha'\beta\beta'} \langle\beta|d_\sigma|\alpha\rangle\langle\alpha'|d_\sigma|\beta'\rangle\langle\langle\alpha\beta|G^{-1}|\alpha'\beta'\rangle\rangle. \tag{8.2}$$

In equation (8.2), α, α' and β, β' mark the Stark sublevels of the upper (a) and lower (b) states, between wich the radiative transition. Further, in equation (8.2), d_σ are projections of the dipole moment operator, and the spectral operator G is

$$G = i\Delta\omega + F_{ab}(v). \tag{8.3}$$

In equation (8.3), $\Phi_{ab}(v)$ is the impact operator that can be represented in the form:

$$\Phi_{ab} = N_b v \int_0^\infty 2\pi\rho d\rho \{S_a S_b^* - 1\}_{\vec{\rho}}. \tag{8.4}$$

In equation (8.4), N_b is the electron density of the REB.

In paper [10], the operator $\Phi_{ab}(v)$ was broken down into adiabatic $\Phi^{ad}{}_{ab}(v)$ and nonadiabatic $\Phi^{na}{}_{ab}(v)$ parts:

$$\Phi_{ab}(v) = \Phi^{ad}_{ab}(v) + \Phi^{na}_{ab}(v). \tag{8.5}$$

In equation (8.5), the adiabatic part $\Phi^{ad}{}_{ab}(v)$ contained only the diagonal matrix elements of the dipole moment operator in the following combination:

$e^2(z_{\alpha\alpha} - z_{\beta\beta})^2$. An important property of the dynamical Stark broadening by an electron beam or by an ion beam beam is that the adiabatic contribution $\Phi^{ad}{}_{ab}(v)$ is zero, what differs from the dynamical Stark broadening by thermal electrons or ions that move randomly [12].

The scattering matrix S in equation (8.4) can be expressed as follows:

$$S = \exp\left[(i/\hbar)\int_{-\infty}^{\infty} dt\ d_z E_z(t)\right]\hat{T}\exp\left[(i/\hbar)\int_{-\infty}^{\infty} dt\ Q^*(d_x E_x + d_y E_y)Q\right],$$
$$Q = \exp\left[-(i/\hbar)(H_0 t - \int_{-\infty}^{t} dt'\ d_z E_z(t'))\right]. \tag{8.6}$$

For the Lyman lines the analytical calculations are simpler because the scattering matrix $S_b = 1$. This simplifies calculations. In this case, in the second order of the modified Dyson expansion (8.6), the nonadiabatic contribution to the operator $\Phi_{\mathrm{ab}}(v)$ has the following matrix element:

$$\Phi_{\alpha\alpha'} = -4\pi\ N_b \frac{e^2}{\hbar^2 v}\sum_{\alpha''} d^x_{\alpha\alpha''}\ d^x_{\alpha''\alpha'}\int_0^{\infty} C_{\pm}(Z)\frac{dZ}{Z}. \tag{8.7}$$

In equation (8.7),

$$Z = 2m_e v\rho/(3n\hbar). \tag{8.8}$$

In equation (8.8), ρ is the impact parameter and n is the principal quantum number of the upper level. From the physical point of view, the parameter Z is the scaled, dimensionless impact parameter. Therefore, the integral over Z in equation (8.7) is actually the integral over impact parameters.

In the case of a nonrelativistic electron beam, it would produce the following electric field at the position of the radiating atom:

$$\mathbf{E}(t) = e\mathbf{r}(t)/r^3(t). \tag{8.9}$$

In equation (8.9), $r(t)$ is the radius vector directed from the electron in the beam to the radiating atom. In this situation, the broadening functions C_- and C_+ in equation (8.7), corresponding to diagonal and nondiagonal matrix elements, respectively, would be as follows:

$$C_{\pm}(Z) = \frac{1}{2}\int_{-\infty}^{\infty}\int_{-\infty}^{x_1}\frac{dx_1 dx_2}{[g(x_1)g(x_2)]^3}\exp\left[\frac{i}{Z}\left(1/g(x_1) \pm 1/g(x_2)\right)\right],$$
$$g(x) \equiv \sqrt{1 + x^2}. \tag{8.10}$$

However, for the case of the REB, equation (8.9) has to be transformed to (see, for example, equation (38.8) from textbook [13]):

$$\mathbf{E}(t) = e\mathbf{r}(t)/\left[r^3(t)\gamma^2(\cos^2\theta + \sin^2\theta/\gamma^2)^{3/2}\right]. \tag{8.11}$$

In equation (8.11),

$$\gamma = 1/\left(1 - v^2/c^2\right)^{1/2}. \tag{8.12}$$

is the so-called relativistic factor. Further, in equation (8.11), $\theta(t)$ is the angle between vector $\boldsymbol{r}(t)$ and the beam velocity v, so that

$$\cos^2\theta = v^2t^2/(\rho^2 + v^2t^2), \ \sin^2\theta = \rho^2/(\rho^2 + v^2t^2). \tag{8.13}$$

In equation (8.13), the instant $t = 0$ was chosen to correspond to the instant of the closest approach of the beam electron to the radiating atom.

The relativistic broadening functions C_{r-} and C_{r+} can be represented in the form:

$$C_{r\pm}(Z) = \frac{1}{2\gamma^4}\int_{-\infty}^{\infty}\int_{-\infty}^{x_1}\frac{dx_1dx_2}{\left[g_r(x_1)g_r(x_2)\right]^3}\exp\left[\frac{i}{Z}\left(1/g_r(x_1) \pm 1/g_r(x_2)\right)\right],$$
$$g_r(x) \equiv \sqrt{1/\gamma^2 + x^2}. \tag{8.14}$$

For the real parts of the relativistic broadening functions, denoted $A_{r\pm} = \mathrm{Re}\ C_{r\pm}$, the double integral in equation (8.14) was calculated in paper [10] analytically, resulting in the following:

$$A_{r-} = (\pi/2)^2\left[\mathbf{H}_{-1}(1/s) + J_1(1/s)\right],\ A_{r+} = (\pi/2)^2\left[\mathbf{H}_{-1}(1/s) - J_1(1/s)\right],\ s = Z/\gamma. \tag{8.15}$$

In equation (8.15), $J_1(1/s)$ and $\mathbf{H}_{-1}(1/s)$ are the Bessel and Struve functions, respectively. Below for brevity, the subscript 'r' is omitted.

The Stark width of the components of hydrogen or deuterium spectral lines is controlled by the subsequent integration over the scaled impact parameter Z:

$$a_\pm = \int_0^{Z\max} A_\pm(Z)dZ/Z = \int_0^{Z\max/\gamma}\int A_\pm(s)ds/s,\ s = Z/\gamma. \tag{8.16}$$

Figure 8.1 displays the dependence of the integrand $A_-(s)/s$ on parameter s. One can see that the corresponding integral a_- converges at small impact parameters—in distinction to the standard, less advanced formalism of the Stark broadening theory plagued by the divergency at small impact parameters.

Figure 8.2 shows the dependence of the integrand $A_+(s)/s$ from equation (8.16) on the parameter s.

Figure 8.3 demonstrates a magnified part of the plot from figure 8.2 at small impact parameters.

From figures 8.2 and 8.3 one can see that the integral a_+ in equation (8.16) also converges at small impact parameters.

At large Z the integrals in equation (8.16) diverge, which is the same situation in the standard theory of the Stark broadening of hydrogen or deuterium spectral lines. The physics behind this divergency is the long-range character of the Coulomb interaction. Fortunately, there is a natural upper cutoff $Z_{\max}$ because of the Debye screening in plasmas:

$$Z_{\max} = uZ_0,\ u = v/c = (1 - 1/\gamma^2)^{1/2}, \quad Z_0 = 2m_ec\rho_D/(3n\hbar). \tag{8.17}$$

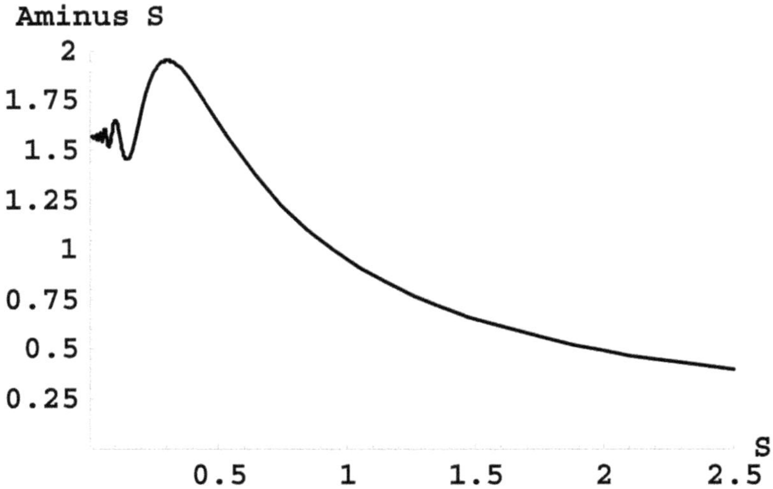

Figure 8.1. Dependence of the integrand $A_{-}(s)/s$ from equation (8.16) that corresponds to the widths function a_{-}, on the parameter $s = Z/\gamma$ defined in equation (8.16). Reprinted from [10]. Copyright IOP Publishing Ltd. CC BY 3.0.

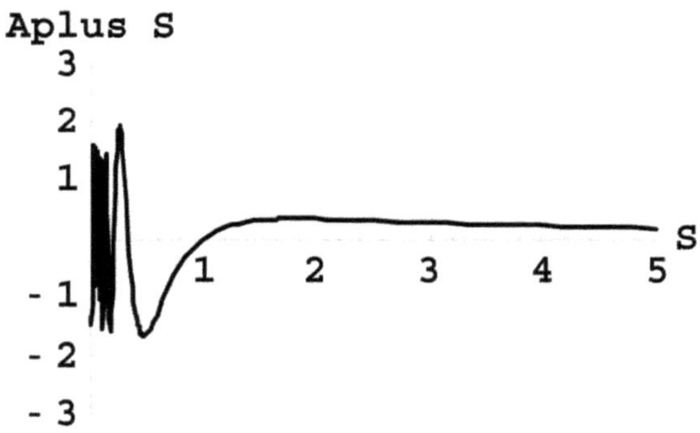

Figure 8.2. Dependence of the integrand $A_{+}(s)/s$ from equation (8.16) that corresponds to the widths function a_{-}, on the parameter $s = Z/\gamma$ defined in equation (8.16). Reprinted from [10]. Copyright IOP Publishing Ltd. CC BY 3.0.

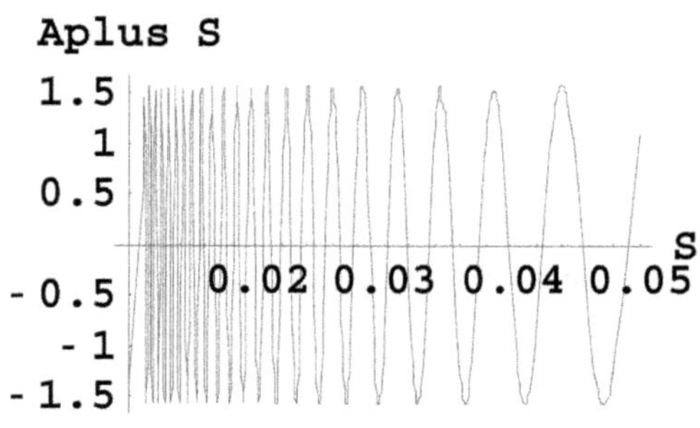

Figure 8.3. Same as in figure 8.2, but for small impact parameters. Reprinted from [10]. Copyright IOP Publishing Ltd. CC BY 3.0.

In equation (8.17),

$$\rho_D = [T_e/(4\pi e^2 N_e)]^{1/2}. \tag{8.18}$$

In equation (8.18), N_e and T_e are the density and the temperature of the bulk plasma electrons, respectively.

The integration in equation (8.16) can be achieved analytically. This is because the integrals in equation (8.16) have the following antiderivatives

$$j_\pm(s) = \int A_\pm(s)ds/s = \left(\pi^2/8\right)\Big\{\left(2/\pi\right)\text{MeijerG}[\{\{0\},\{1\}\},\{\{0,0\},\{-1/2,1/2\}\},1/\left(4s^2\right] + \mathbf{H}_{-1}{}^2(1/s) + \mathbf{H}_0{}^2(1/s) \pm \left[1 - {}_1F_2\left(1/2;1,2;-1/s^2\right)\right]\Big\}. \tag{8.19}$$

In equation (8.19), ${}_1F_2(\ldots)$ and MeijerG[…] are the generalized hypergeometric function and the MeijerG function, respectively. In this way, the authors of paper [10] produced the following analytical results for the width functions:

$$a_\pm = j_\pm\left(Z_{\max}/\gamma\right) - j_\pm(0). \tag{8.20}$$

As an example, the authors of paper [10] calculate in the explicit form the profile I $(\Delta\omega,\gamma)$ of the Lyman-alpha spectral line broadened by an REB, $\Delta\omega$ being the detuning from the unperturbed frequency of the spectral line. Analogously to paper [12], they inverted the spectral operator analytically to obtain:

$$I(\Delta\omega,\ \gamma) = \frac{1}{3\pi}\left(\frac{\Gamma_\pi}{\Delta\omega^2 + \Gamma_\pi^2} + \frac{2\,\Gamma_\sigma}{\Delta\omega^2 + \Gamma_\sigma^2}\right), \tag{8.21}$$

In equation (8.21), Γ_σ and Γ_π are the half-widths at half-maximum of the σ- and π-components of the Lyman-alpha line, respectively, for which the following expressions were obtained in paper [10]:

$$\Gamma_\sigma = \left[\eta_0/(1 - 1/\gamma^2)^{1/2}\right]\left[j_-(Z_{\max}/\gamma) - j_-(0)\right], \tag{8.22}$$

$$\Gamma_\pi = \left[\eta_0/(1 - 1/\gamma^2)^{1/2}\right]\int_0^\infty \left[A_-(s) - A_+(s)\right]ds/s. \tag{8.23}$$

In equations (8.22) and (8.23),

$$\eta_0 = 4\pi\hbar^2 N_e/(m_e{}^2 c) = 5.618 \times 10^{-10} N_e(\text{cm}^{-3})\text{s}^{-1}. \tag{8.24}$$

It should be noted that in equation (8.23), the upper limit of the integration is set as infinity. The reason is that for the π-component of the Lyman-alpha line, the Stark width in equation (8.23) is controlled by the difference of diagonal and nondiagonal matrix elements of the broadening operator. Because of this fortunate fact, the corresponding integration converges both at small and large impact parameters. As a result, the following relatively simple formula was obtained in paper [10] for the Stark width of the π-component of the Lyman-alpha line:

$$\Gamma_\pi = \pi^2\eta_0/[4(1 - 1/\gamma^2)^{1/2}]. \tag{8.25}$$

Figure 8.4 displays the dependence of the scaled width of the σ-component Γ_σ/η_0 (upper curve) and of the scaled width of the π-component Γ_π/η_0 (lower curve) of the Lyman-alpha line broadened by an REB on the relativistic factor γ at the bulk plasma electron density $N_e = 10^{15}$ cm^{-3} and the bulk plasma electron temperature $T_e = 2$ eV. One can see that as γ increases from unity, both Stark widths significantly diminish.

Figure 8.5 shows the dependence of the ratio Γ_σ/Γ_π versus the relativistic factor γ at the bulk plasma electron density $N_e = 10^{15}$ cm^{-3} and the bulk plasma electron temperature $T_e = 2$ eV. One can see that as γ increases, this ratio initially increases, then reaches the maximum, and then diminishes. The maximum ratio, reached at $\gamma = 2^{1/2}$, is equal to $\Gamma_\sigma/\Gamma_\pi = 5.39$.

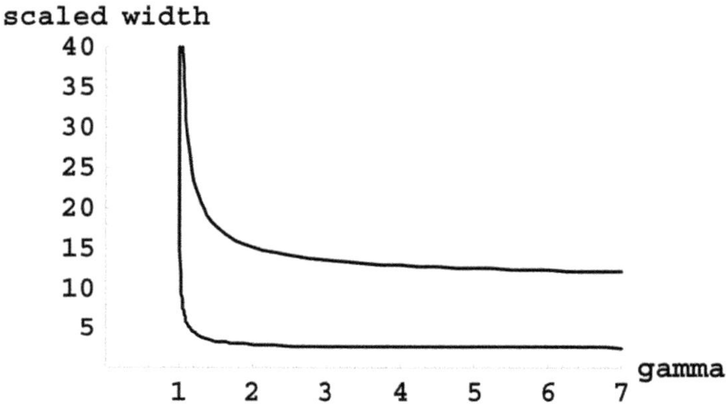

Figure 8.4. Dependence of the scaled width of the σ-component Γ_σ/η_0 (upper curve) and the scaled width of the π-component Γ_π/η_0 (lower curve) of the hydrogen or deuterium Lyman-alpha line broadened by a REB on the relativistic factor γ at the bulk plasma electron density $N_e = 10^{15}$ cm^{-3} and the bulk plasma electron temperature $T_e = 2$ eV. Reprinted from [10]. Copyright IOP Publishing Ltd. CC BY 3.0.

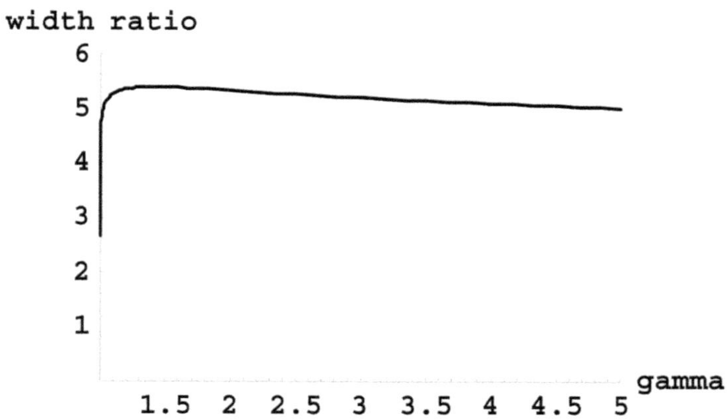

Figure 8.5. Dependence of the ratio Γ_σ/Γ_π versus the relativistic factor γ at the bulk plasma electron density $N_e = 10^{15}$ cm^{-3} and the bulk plasma electron temperature $T_e = 2$ eV. Reprinted from [10]. Copyright IOP Publishing Ltd. CC BY 3.0.

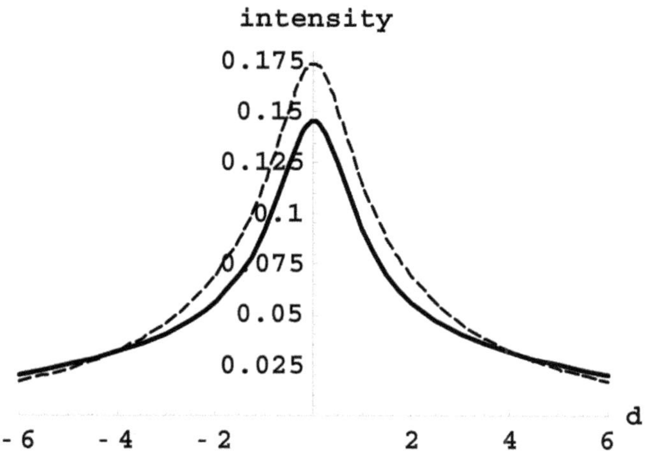

Figure 8.6. Calculated unpolarized profiles of the entire Lyman-alpha line for the direction of the observation orthogonal to the velocity of the REB for the bulk plasma parameters $N_e = 10^{15}$ cm^{-3} and $T_e = 2$ eV. The profiles are displayed versus the scaled detuning $\Delta\omega/\Gamma_\pi$ denoted as d. The solid line corresponds to the parameter $\gamma = 2^{1/2}$, the dashed line—to $\gamma = 1$. Reprinted from [10]. Copyright IOP Publishing Ltd. CC BY 3.0.

By introducing a polarizer in the optical system and recording the profile of the same spectral line in two orthogonal polarizations—one polarization parallel to the REB velocity and the other polarization perpendicular to the REB velocity—it is possible to measure Γ_π or Γ_σ, respectively, and thus to determine the experimental ratio Γ_σ/Γ_π. By monitoring the time evolution of this experimental ratio, one should be able to detect the appearance of an REB in tokamaks and thus to initiate—in the timely fashion—the mitigation of the problem.

Figure 8.6 displays the calculated unpolarized profiles of the entire Lyman-alpha line for the case where the direction of the observation is orthogonal to the velocity of the REB for the following bulk plasma parameters: $N_e = 10^{15}$ cm^{-3} and $T_e = 2$ eV. The profiles are presented as the function of the scaled detuning $\Delta\omega/\Gamma_\pi$, which is denoted as d. This scaling makes the calculated profiles universal: they do not depend on the electron density of the beam. The profiles correspond to two values of the parameter γ, namely $\gamma = 2^{1/2}$ (solid line) and $\gamma = 1$ (dashed line).

For the value of $\gamma = 2^{1/2}$ the profile is narrower (by 12%) than for the case of $\gamma = 1$. It would be much more difficult to detect the appearance of an REB by monitoring this relatively small decrease of the width. Therefore, the usage of the polarization analysis for monitoring the ratio Γ_σ/Γ_π, which would increase by an order of magnitude during the development of an REB in the plasma, is more promising.

8.2 The case of the high energy hydrogen or deuterium beam

The design of the future tokamak ITER contains a deuterium beam of the energy $E = 1$ MeV and a hydrogen beam of energy $E = 0.87$ MeV [14]. In the reference frame of the beam atoms, their anisotropic broadening of the spectral lines of the beam atoms is due to the combined action of the magnetic field **B** and the Lorentz field $\mathbf{F} = \mathbf{v} \cdot \mathbf{B}/c$. The fields **F** and **B** are orthogonal to each other, the field **F** having the absolute value $\mathbf{vB}(\sin\theta)/c$, where θ is the angle between vectors **v** and **B**.

Here we consider the situation where the beam velocity

$$v = \left[E/(gM_p)\right]^{1/2}, \tag{8.26}$$

significantly exceeds the thermal velocity of the plasma electrons $v_{Te} = (2T_e/m_e)^{1/2}$:

$$v/v_{Te} = \left\{\left[m_e/(gM_p)\right]E/T_e\right\}^{1/2} > 1. \tag{8.27}$$

In equations (8.26) and (8.27), E is the beam energy, $g = 1$ for the hydrogen beam or $g = 2$ for the deuterium beam, and M_p is the proton mass. In terms of the beam energy, the condition (8.27) means

$$E/T_e > gM_p/m_e = 1.84 \times 10^3\, g. \tag{8.28}$$

For instance, for ITER the planned hydrogen beam will have the energy 870 keV [14]. Since at the edge plasma T_e 10 eV, then the condition (8.28) becomes:

$$E/T_e \geqslant 8.7 \times 10^4 \gg 1.84 \times 10^3. \tag{8.29}$$

For the planned deuterium beam of the energy 1 MeV [14], planned for ITER [5], the condition (8.28) becomes:

$$E/T_e \geqslant 10^5 \gg 1.84 \times 10^3\, g = 3.68 \times 10^3. \tag{8.30}$$

Therefore, in this case, in the reference frame of the beam atoms, the velocity of the plasma electrons and of the plasma ions can be considered as $-\mathbf{v}$ with a good accuracy.

In the crossed electric and magnetic fields, the separation between the adjacent sublevels of the split levels of the beam atoms (in the frequency scale) is as follows (according to paper [15] by Demkov, Monozon, and Ostrovskii, hereafter, DMO):

$$\Delta\omega_{\mathrm{DMO}} = \left\{\left[3n\hbar F/(2m_e e)\right]^2 + (\mu_{\mathrm{B}}B/\hbar)^2\right\}^{1/2}. \tag{8.31}$$

In equation (8.31), μ_{B} is the Bohr magneton. For the nonadiabatic virtual transitions between these sublevels to occur, the maximum frequency of the variation of the electric field v/ρ of the plasma electrons or plasma ions must be greater than the separation between the sublevels

$$v/\rho > \Delta\omega_{\mathrm{DMO}}. \tag{8.32}$$

The corresponding restriction on the impact parameter is:

$$\rho < v/\Delta\omega_{\mathrm{DMO}} = \rho_{\mathrm{max}}. \tag{8.33}$$

Since in our case, $F = vB\sin\theta$, then

$$\rho_{\mathrm{max}} = [vm_e c/(eB)]/\{[(3n/2)(\hbar c/e^2)(v/c)\sin\theta]^2 + 1\}^{1/2}. \tag{8.34}$$

In equation (8.34), the first term under the square root is the square of the ratio of the Stark splitting $\Delta\omega_{Stark}$ in the Lorentz field to the Zeeman splitting $\Delta\omega_{Zeem}$ in the magnetic field. The former significantly exceeds the latter if

$$v/c \gg [e^2/(\hbar c)]2/(3n \sin\theta) = 1/(205.6n \sin\theta). \tag{8.35}$$

In terms of the beam energy, the condition (8.35) means:

$$E(\text{MeV}) \gg 0.0111g/(n^2 \sin^2\theta) = E_{crit}. \tag{8.36}$$

The cosine of the planned angle θ for ITER is about 0.3 [16]; consequently, $\sin^2\theta \approx 0.9$. For the hydrogen beam of the energy $E = 0.87$ MeV, the condition (8.36) becomes

$$E/E_{crit} \approx 70n^2 \geqslant 280, \tag{8.37}$$

since $n \geqslant 2$. For the deuterium beam of the energy $E = 1$ MeV, the condition (8.36) becomes

$$E/E_{crit} \approx 40n^2 \geqslant 160. \tag{8.38}$$

One can see that for the atoms of the above hydrogen or deuterium beams, the Stark splitting preponderates over the Zeeman splitting. Consequently, equation (8.34) can be simplified as follows:

$$\rho_{max} \approx 2m_e ec/(3n\hbar B \sin\theta). \tag{8.39}$$

It is worth emphasizing that in equation (8.39), ρ_{max} is independent of the beam velocity.

The next step is to compare the Debye radius $\rho_{De} = [T_e/(4\pi e^2 N_e)]^{1/2}$ to ρ_{max} from equation (8.39):

$$\begin{aligned}\rho_{De}^2/\rho_{max}^2 &\approx 9\hbar^2 n^2 T_e B^2 \sin^2\theta/(16\pi e^4 m_e^2 c^2 N_e) = \\ &8.03 \times 10^3 n^2 T_e(\text{eV})\left[B(\text{Gauss})\right]^2(\sin^2\theta)/N_e(\text{cm}^{-3}).\end{aligned} \tag{8.40}$$

At the edge of ITER, the expected plasma parameters are as follows: $N_e = (1-3) \cdot 10^{14}$ cm^{-3} (see, e.g., [17, 18]), $T_e = 10$ eV (see, e.g., [19]), and $B \approx 4$ Tesla (see, e.g., [17]), $T_e = 10$ eV (see, e.g., [19]). Then by utilizing $\sin^2\theta \approx 0.9$, we obtain

$$\rho_{De}^2/\rho_{max}^2 \approx (0.390 - 1.17)n^2 \geqslant (1.6 - 4.7), \tag{8.41}$$

since $n \geqslant 2$. This means that the integration over the impact parameters (in calculating the Stark width) must be truncated at ρ_{max}—in distinction to truncating at ρ_{De}, as was done in paper [12]).

In terms of the scaled impact parameter

$$Z(\rho) = 2m_e vr/(3n\hbar), \tag{8.42}$$

ρ_{max} corresponds to

$$Z_{max} = 4m_e^2 ecv/(9n^2\hbar^2 B \sin\theta) = 6.62 \times 10^6[E(\text{MeV})/g]^{1/2}/[n^2 \sin\theta B(\text{Tesla})]. \tag{8.43}$$

Several points should be emphasized. First, according to equation (8.43), the value $Z_{\max}$ is the same both for the anisotropic broadening by plasma electrons and for the anisotropic broadening by plasma ions. Second, for plasmas where the effective ion charge is approximately unity, in the formula for the impact broadening operator

$$\Phi_{\alpha\alpha'} = -[4\pi N_p Z_p{}^2 e^2/(\hbar^2 v)]\sum_{\alpha''}(d_x)_{\alpha\alpha''}(d_x)_{\alpha''\alpha'}c_{\pm} \tag{8.44}$$

(where N_p is the density of perturbers, Z_p is the charge of the perturbers, and $c_{\pm}$ are the broadening functions, the latter depending on the cutoff $Z_{\max}$ from equation (8.43)), the factor $4\pi N_p Z_p{}^2 e^2/(\hbar^2 v)$ is the same both for plasma electrons and for plasma ions. Consequently, in this case, both the plasma electrons and the plasma ions provide the *equal contributions* to the Stark broadening, which is equivalent to *doubling the density* N in equation (8.44). This is a *counterintuitive result.*

There is also another interesting property of the above anisotropic Stark broadening. Namely, it *does not have any dependence on the reduced mass* of the 'radiator–perturber' pair. This is a clear distinction from the situation with the isotropic Stark broadening where the result depended on the reduced mass of the pair 'perturber—radiator'—the dependence that was important for the dynamical Stark broadening by plasma ions.

Yet another interesting property of the above anisotropic Stark broadening is the following. It is *exactly linear* with respect to the electron density N_e and/or to the ion density N_i. This is yet another clear distinction from the isotropic Stark broadening, the latter being proportional to $N_{e,i}\ln[f(T,n)/N_{e,i}]$, where $f(T,n)$ is some function of T (the plasma temperature) and of n (the principal quantum number). Such nonlinear dependence was the case for all the previous theories of the isotropic Stark broadening. Moreover, even in the case of the anisotropic Stark broadening studied in papers [12, 20], the broadening was still proportional to $N_{e,i}\ln[f(T,n)/N_{e,i}]$. Thus, the exactly linear dependence of the anisotropic Stark broadening on N_e and/or N_i presented here is *yet another counterintuitive result.*

For an *arbitrary* Lyman line of hydrogen or deuterium, the real part of the diagonal matrix elements of the impact operator is

$$Re\Phi_{\alpha\alpha} = -[8\pi N_e e^2/(\hbar^2 v)]\sum_{\alpha''}(d_x)_{\alpha\alpha''}(d_x)_{\alpha''\alpha}[\ln Z_{\max} + \pi^{5/2}/12 + (32 - 3\pi^2)/(96 Z_{\max}{}^2)]. \tag{8.45}$$

The corresponding nondiagonal matrix elements are as follows

$$Re\Phi_{\alpha\alpha'} = -[8\pi N_e e^2/(\hbar^2 v)]\sum_{\alpha''}(d_x)_{\alpha\alpha''}(d_x)_{\alpha''\alpha'}[\ln Z_{\max} - 1 + (32 + 3\pi^2)/(96 Z_{\max}{}^2)],\ \alpha' \neq \alpha, \tag{8.46}$$

where $Z_{\max}(v, n, B, \theta)$ was defined in equation (8.43). For the ITER 0.87 MeV hydrogen beam at the magnetic field $B = 4$ Tesla and the angle $\theta = 73°$ (corresponding to $\cos\theta = 0.3$):

$$Z_{\max} = 1.62 \times 10^6/n^2. \tag{8.47}$$

For the ITER 1 MeV deuterium beam at the same plasma parameters;

$$Z_{\max} = 1.23 \times 10^6/n^2. \tag{8.48}$$

It is important to emphasize that in the case of the isolated components of the spectral line, the real part of the diagonal matrix elements $Re\Phi_{\alpha\alpha}$ is equal (with the minus sign) to the Stark half-width at half-maximum γ_α of the spectral component:

$$\gamma_\alpha = -Re\Phi_{\alpha\alpha}. \tag{8.49}$$

For the Lyman-alpha line, more explicit analytical results are presented below. For the central Stark component (the σ-component):

$$\gamma_\sigma = \left[8\pi N_e\hbar^2/\left(m_e{}^2 v\right)\right]\left[\ln Z_{\max} + \pi^{5/2}/12 + \left(32 - 3\pi^2\right)/\left(96Z_{\max}{}^2\right)\right]. \tag{8.50}$$

For the lateral Stark components (the π-components):

$$\gamma_\pi = \left[8\pi N_e\hbar^2/\left(m_e{}^2 v\right)\right]\left[\pi^2/4 - \pi^2/\left(16Z_{\max}{}^2\right)\right]. \tag{8.51}$$

For further details see paper [21].

References

[1] Guenot D *et al* 2017 *Nat. Photon.* **11** 293
[2] Kurkin S A, Hramov A E and Koronovskii A A 2013 *Appl. Phys. Lett.* **103** 043507
[3] de Jagher P C, Sluijter F W and Hopman H J 1988 *Phys. Rep.* **167** 177
[4] Decker J, Hirvijoki E, Embreus O, Peysson Y, Stahl A, Pusztai I and Fülöp T 2016 *Plasma Phys. Control. Fusion* **58** 025016
[5] Smith H, Helander P, Eriksson L-G, Anderson D, Lisak M and Andersson F 2006 *Phys. Plasmas* **13** 102502
[6] Minashin P V, Kukushkin A B and Poznyak V I 2012 *EPJ Web Conf.* **32** 01015
[7] Kurzan B, Steuer K-H and Suttrop W 1997 *Rev. Sci. Instrum.* **68** 423
[8] Ide S *et al* 1989 *Nucl. Fusion* **29** 1325
[9] Oks E 2017 *Diagnostics of Laboratory and Astrophysical Plasmas Using Spectral Lines of One-, Two-, and Three-Electron Systems* (Hackensack, NJ: World Scientific)
[10] Oks E and Sanders P 2018 *J. Phys. Commun.* **2** 015030
[11] Rosato J, Pandya S P, Logeais C, Meireni M, Hannachi I, Reichle R, Barnsley R, Marandet Y and Stamm R 2017 *AIP Conf. Proc.* **1811** 110001
[12] Derevianko A and Oks E 1995 *J. Quant. Spectrosc. Radiat. Transfer* **54** 137
[13] Landau L D and Lifshitz E M 1971 *The Classical Theory of Fields* (Oxford: Pergamon)
[14] Hemsworth A S *et al* 2017 *New J. Phys.* **19** 025005
[15] Demkov Y, Monozon B and Ostrovskii V 1970 *Sov. Phys. JETP* **30** 775
[16] Kukushkin A B 2023 Private communication.
[17] Watts C *et al* 2013 *Nucl. Instrum. Methods Phys. Res.* A **720** 7
[18] Yatsuka E *et al* 2013 *J. Instrum.* **8** C12001
[19] Kukushkin A B, Kukushkin A S, Lisitsa V S, Neverov V S, Pshenov A A and Shurygin V A 2021 *Plasma Phys. Control. Fusion* **63** 035025
[20] Derevianko A and Oks E 1997 *Rev. Sci. Instrum.* **68** 998
[21] Oks E 2024 *IEEE Trans. on Plasma Sci.* (accepted)

Chapter 9

Polarization and directional effects in the Stark broadening of spectral lines in strongly-magnetized plasmas

9.1 Remarkable polarization properties of the Stark profiles and of the Stark–Zeeman profiles of the hydrogen Lyman-alpha line

A very important parameter for operations of tokamaks is the pitch angle

$$\gamma_p = \tan^{-1}\!\left(B_P/B_T\right), \tag{9.1}$$

where B_P and B_T are the poloidal and toroidal magnetic fields, respectively. Measurements of the pitch angle by means of the beam spectroscopy are commonly used—see, e.g., work [1].

Another important parameter for operations of tokamaks is the effective charge of plasma ions

$$Z_{\text{eff}} = \Sigma Z_i^2 N_i / N_e, \tag{9.2}$$

where Z_i and N_i are the charge and the density if the ith sort of ions, while N_e is the electron density. A spectroscopic method measuring the effective charge was proposed suggested in paper [2]. Thereafter, it was implemented in paper [3]. Then it was revitalized in paper [4].

The method was based on the following idea. For plasma parameters characteristic of tokamaks (the electron density $N_e \sim 10^{14}$ cm^{-3}, the hydrogen or deuterium atom temperature $T_a \geqslant 10^2$ eV), the Stark width of hydrogen or deuterium spectral lines is dominated by the ion dynamical broadening. Therefore, the Stark width γ_s, being proportional to $\Sigma Z_i^2 N_i / \langle V_i \rangle$), is linear with respect to the effective charge Z_{eff}.

In paper [4] it was demonstrated that only the Lyman-alpha line can be used for the experimental determination of the effective charge: specifically, only its π-component which in the strong magnetic field is the central, unshifted component of the Lyman-alpha line. This result was obtained by using the advanced formalism [5, 6], as opposed to the conventional theory of the dynamical Stark broadening [7].

Then in paper [8], which we follow here, the authors significantly enhanced the beam spectroscopy diagnostic method. The enhancement enable measuring simultaneously both important parameters in tokamaks: the pitch angle and the effective charge. Below are some brief details.

In paper [8] there was analyzed the situation where a hydrogen or deuterium beam of the velocity **v** enters—at some angle θ—a region of a static uniform magnetic field **B**. The following two conditions are assumed to be satisfied. The first condition is:

$$v \ll v_{Te}, \tag{9.3}$$

where v_{Te} is the electron mean thermal velocity. Under this condition, dynamical Stark broadening of the beam atoms by plasma electrons can be disregarded compared to the dynamical Stark broadening by plasma ions.

The second condition is:

$$v \ll v_0 = 2e^2/(3n\hbar \sin\theta) = c/(205.5n \sin\theta). \tag{9.4}$$

In equation (9.4), n is the principal quantum number of the corresponding hydrogenic energy level. For the beam energy E, equation (9.4) is equivalent to the following:

$$E(\text{keV}) \ll E_0(\text{keV}) = 11.12(M/M_H)/(n \sin\theta)^2. \tag{9.5}$$

In equation (9.5), M is the mass of the beam atoms, while M_H is the mass of hydrogen atoms. Under the condition (9.4) or (9.5), the Zeeman splitting in the field **B** predominates over the Stark splitting in the Lorentz field $\mathbf{v} \cdot \mathbf{B}/c$.

After analytical calculations similar (but not identical) to those presented in paper [4], the authors of paper [8], first of all, arrived at the conclusion that out of all components of all hydrogenic lines, only the π-component (that is, the central, unshifted component) of the Lyman-alpha line can be used for measuring the effective charge Z_{eff}. Second, in paper [8] it was shown that the dynamical Stark width (the half-width at half-maximum) of the π-component depends on the angle θ, as follows:

$$\gamma_\pi(s^{-1}) = 2.17 \times 10^{-4} Z_{\text{eff}} \left\{ M/\left[M_H E(\text{eV}) \right] \right\}^{1/2} N_e(\text{cm}^{-3}) f(R) \sin^2\theta. \tag{9.6}$$

In equation (9.6),

$$R = 1.49 \times 10^8 Z_{\text{eff}}{}^{-3/2} \left\{ E(eV) T(eV) M_H / \left[N_e(\text{cm}^{-3}) M \right] \right\}^{1/2} / \sin\theta, \tag{9.7}$$

and

$$f(R) \approx 0.279 + (\ln R)/4. \tag{9.8}$$

In the geometry usually employed in the experimental beam spectroscopy in tokamaks, the angle α between the toroidal magnetic field $\mathbf{B}_T$ at the point of observation and the beam velocity $\mathbf{v}$ is known. The beam propagates in the tokamak horizontal mid-plane—therefore, the poloidal magnetic field $\mathbf{B}_P$ is directed vertically. Thus, the angle θ in equations (9.6) and (9.7), which is the angle between the vectors $\mathbf{B} = \mathbf{B}_T + \mathbf{B}_P$ and $\mathbf{v}$, is related to angle α and to the pitch angle as follows:

$$\cos\gamma_p = \cos\theta/\cos\alpha. \tag{9.9}$$

The angle Ω between the wave vector $\mathbf{k}$ of radiated photons (which is the direction of the observation) and the vector $\mathbf{B}_T$ is also known. Typically, the vector $\mathbf{k}$ is exactly or approximately in the mid-plane. Therefore, the angle γ_m between the vectors $\mathbf{B} = \mathbf{B}_T + \mathbf{B}_P$ and $\mathbf{v}$ (the tilt angle) is related to the angle Ω and to the pitch angle as follows:

$$\cos\gamma_p = \cos\gamma_m/\cos\Omega. \tag{9.10}$$

In paper [8] there were proposed two different techniques for measuring simultaneously the effective charge Z_{eff} and the pitch angle γ_p. The first technique should employ a linear polarizer in the optical system. One should measure frequency-integrated intensity I_π of the Lyman-alpha π-component in two mutually perpendicular polarizations. Namely, the intensity

$$I_{\pi1} = I_{\max}\sin^2\gamma_p, \tag{9.11}$$

corresponding to the vertical orientation of the polarizer, and the intensity

$$I_{\pi2} = I_{\max}\sin^2\Omega\cos^2\gamma_p \tag{9.12}$$

in the polarization perpendicular to the previous one. Then from the experimental ratio $I_{\pi1}/I_{\pi2}$ one can determine the pitch angle as follows:

$$\tan^2\gamma_p = (I_{\pi1}/I_{\pi2})\sin^2\Omega. \tag{9.13}$$

Then one should calculate the value of $\sin^2\theta$, which according to equation (9.9) is:

$$\sin^2\theta = 1 - \cos^2\alpha\cos^2\gamma_p. \tag{9.14}$$

Then, one should measure the dynamical Stark width of the Lyman-alpha π-component and determine from it the effective charge Z_{eff} by utilizing equations (9.6)–(9.8). Finally, one can deduce the pitch angle γ_p from the value of the angle θ by employing equation (9.9)

In the second technique proposed in paper [8], first one should measure the dynamical Stark width $\gamma_{\pi0}$ of the Lyman-alpha π-component without the presence of the neutral beam. Then, by utilizing equations (17) and (18) from paper [4], one can deduce the effective charge from the measured $\gamma_{\pi0}$. Next, one should measure the dynamical Stark width γ_π of the Lyman-alpha π-component at the presence of the neutral beam. Then from the ratio $\gamma_\pi/\gamma_{\pi0}$, one can determine the value of $\sin^2\theta$ by employing equations (9.6)–(9.8) of this chapter. Finally, the pitch angle γ_p can be deduced from the value of the angle θ by utilizing equation (9.9).

Another remarkable feature of the π-component of the Lyman-alpha line was revealed in paper [9]. Under consideration was the case of the dominance of the Zeeman effect over the Stark effects. It was shown that for the π-component of the Lyman-alpha line, its broadening is essentially controlled by the Stark effect: practically no dependence on the magnetic field. This was a clear distinction from the σ-component of the Lyman-alpha line or from any component of any other spectral line of hydrogenic atoms or ions. Below are some details.

We considered the combined effect of a magnetic field **B** and an electric field **F** at the angle θ with respect to **B** on the Lyman-alpha line of hydrogenic atoms or ions. The following notations were introduced

$$M = \alpha B/2, \quad E = 3F/Z, \tag{9.15}$$

where α was the fine structure constant; 'E' stood for 'electric', while 'M' stood for 'magnetic' (atomic units $\hbar = m_e = e = 1$ were utilized).

The eigenvalues of the interaction matrix were the solutions $\Delta\omega$ of the following equation

$$(\Delta\omega)^2 = \left(E^2 + M^2\right)/2 \pm \left[\left(E^2 + M^2\right)^2/4 - E^2M^2\cos^2\theta\right]^{1/2}. \tag{9.16}$$

For the most interesting case of the domination of the magnetic field, that is, for the case where $E \quad M$, the solutions of the equation (9.16) were

$$\Delta\omega \approx \pm[B + (E^2 \sin^2\theta)/2B] \tag{9.17}$$

for the σ-component and

$$\Delta\omega \approx \pm \mathrm{E}\cos\theta \tag{9.18}$$

for the π-component.

From equation (9.18) it is seen that the splitting (or the broadening) of the π-component of the Lyman-alpha line of hydrogenic atoms or ions is linear with respect to the projection of the electric field on the magnetic field. For this reason, the Stark width of this component can be employed in this case either for measuring the root-mean-square field of a Low-frequency Electrostatic Plasma Turbulence (LEPT), (examples of which are ion-acoustic waves, or Bernstein modes, or low-hybrid waves), or for measuring the ion density.

At the fixed magnitude E of the scaled electric field, the Stark profile of the π-component is:

$$S(\Delta\omega, E) = \int_0^1 d(\cos\theta)\delta(|\Delta\omega| - \mathrm{E}\cos\theta) = (1/E)\Theta(E - |\Delta\omega|). \tag{9.19}$$

where $\Theta(\ldots)$ is the theta-function and $\delta(\ldots)$ is the delta-function.

After calculating the average over the Rayleigh distribution of the absolute value of the LEPT, the Rayleigh distribution being (see works [10–12]),

$$W(f)df = 3(6/\pi)^{1/2} f^2 \exp(-3f^2/2)df, \quad f = F/F_t, \tag{9.20}$$

and taking into account that

$$1/E = Z/(3F) = Z/(3F_t f), \tag{9.21}$$

we obtained the following intermediate result for Stark profile of the Lyman-alpha π-component:

$$S(\Delta\omega) = [Z/(3F_t] \int_{f_{\min}}^{\infty} dfW(f)/f, \quad f_{\min}(\Delta\omega) = Z|\Delta\omega|/(3F_t). \tag{9.22}$$

The analytical calculation of the integral in equation (9.22) yielded the following final result:

$$S(\Delta\omega) = [Z/(3F_t] (6/\pi)^{1/2} \exp\{-3 \left[f_{\min}(\Delta\omega)\right]^2/2\}. \tag{9.23}$$

9.2 Polarization and directional effects in the Lorentz–Doppler profiles of hydrogen or deuterium lines

In strongly-magnetized plasmas, such as, for example, in Sun spots, and in the atmospheres of white dwarfs, and in magnetic fusion devices, as hydrogen or deuterium atoms cross the magnetic field **B** with the velocity **v**, they get subjected to the Lorentz electric field $\mathbf{E}_L = \mathbf{v} \cdot \mathbf{B}/c$—in addition to other electric fields in plasmas. Since the atomic velocity **v** has a distribution, then the Lorenz field has a distribution. This is the source of the additional broadening of the spectral lines of these hydrogen or deuterium atoms.

In paper [13] we studied situations where the Lorentz broadening was the primary source of the broadening of Highly-excited Hydrogen/deuterium Spectral Lines (HHSLs). In laboratory plasmas, HHSLs are employed as the tool to measure the electron density at the edge plasmas of tokamaks (see, e.g., papers [14, 15] and section 4.3 of review [16]), as well as in radiofrequency discharges (see, e.g., paper [17] and book [6]). The same situation is also for HHSLs radiated from the solar chromosphere. The corresponding observations are utilized for the measurements of the electron density in the solar chromosphere (see, e.g., paper [18]).

The central point is that in all these cases, combining Lorentz and Doppler broadenings via the convolution is inappropriate, as clarified in paper [19]. These two broadening mechanisms are not independent in these situations. This is because the Lorentz–Doppler profile of a Stark component of HHSLs in the frequency scale is proportional (in the laboratory reference frame) to the following delta-function δ $[\Delta\omega - (\omega_0 v/c)\cos\alpha - (kX_{\alpha\beta}Bv/c)\sin\vartheta]$, where in its argument ϑ is the angle between vectors **v** and **B**, while α is the angle between the direction of observation and the atomic velocity **v**.

In paper [13], for the arbitrary strength of the magnetic field **B** and for the arbitrary angle of the observation ψ with respect **B**, we obtained a general formula for the Lorentz–Doppler profiles of HHSLs. Yet more *specific* analytical results were derived in paper [9] only for the following two values of the angle of the observation: for $\psi = 0°$ and $\psi = 90°$. We demonstrated a substantial *suppression of* π*-components* by a relatively strong magnetic field—compared to σ-components—for the observation angle $\psi = 90°$. This was a *counterintuitive result.*

Then in paper [20] we produced the corresponding *specific* analytical results for an *arbitrary angle of the observation* ψ and for the arbitrary strength of the magnetic field **B**. Below are some details.

Figure 9.1 shows the geometry of the observation relative to the vectors involved.

In paper [13] we introduced the following dimensionless notations:

$$w = c\Delta\omega/vT\omega0 = c\Delta\lambda/vT\lambda0,\ \mathbf{b} = kX\alpha\beta\mathbf{B}/\omega0,\ \mathbf{u} = \mathbf{v}/vT \tag{9.24}$$

In equation (9.24), u is the atomic velocity scaled with respect to the atomic mean thermal velocity v_T, **b** is the scaled magnetic field, and w is the scaled detuning from the unperturbed frequency of a hydrogen spectral line. The other notations in equation (9.24) are

$$k = 3\hbar/(2\mathrm{mee}), \quad X\alpha\beta = n\alpha(n1 - n2)\alpha - n\beta(n1 - n2)\beta. \tag{9.25}$$

In equation (9.25), n is the principal quantum number, while n_1, n_2 are the parabolic quantum numbers; the subscripts α and β relate to the upper and lower Stark sublevels, respectively.

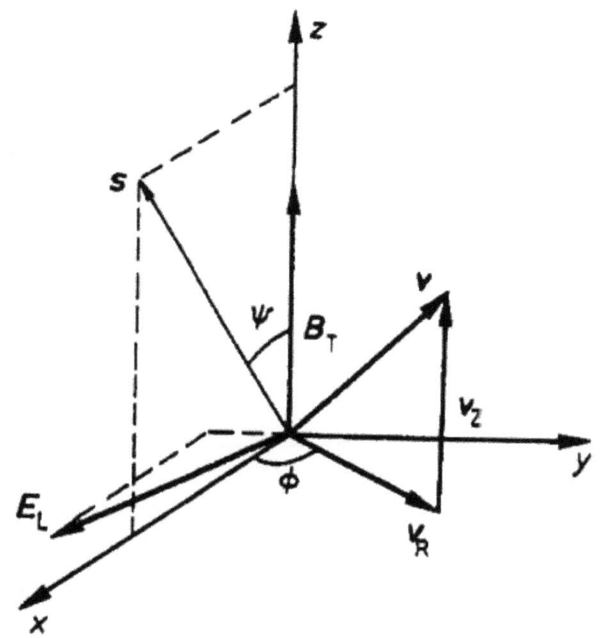

Figure 9.1. Geometry of the observation. The direction of the observation is denoted **s**. The z axis is parallel to **B**. Vector **s** is at the angle ψ to **B**. The atomic velocity **v** has a component $\mathbf{v}_R$ perpendicular to **B** and a component v_z along **B**. The component $\mathbf{v}_R$ is at the angle ϕ to the x axis. Reproduced from [20]. Copyright IOP Publishing Ltd. CC BY 3.0.

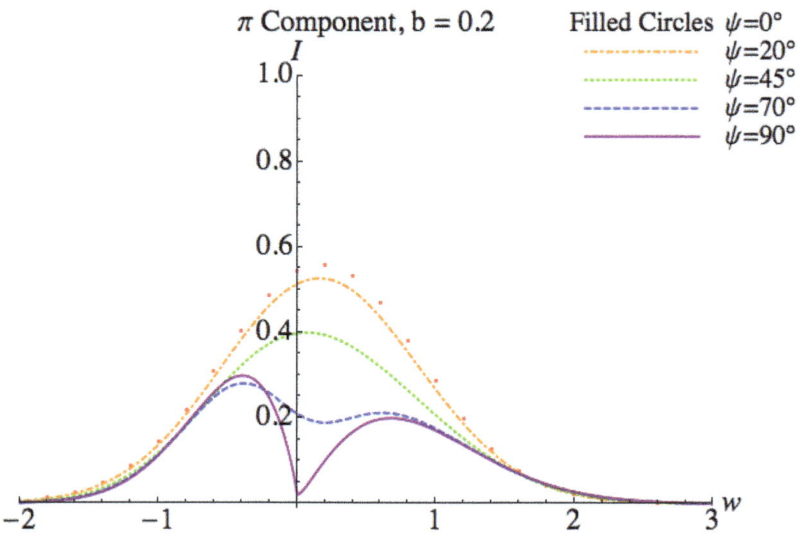

Figure 9.2. Calculated Lorentz–Doppler profiles of π-components of highly-excited hydrogen or deuterium spectral lines for the scaled magnetic field $b = 0.2$ at five different angles of the observation. Reproducd from [20]. Copyright IOP Publishing Ltd. CC BY 3.0.

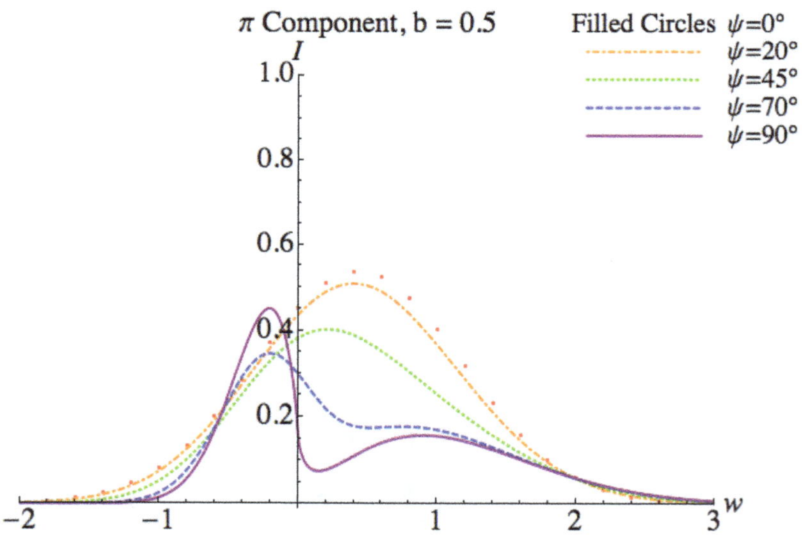

Figure 9.3. Same as in figure 9.2, but for $b = 0.5$. Reproduced from [20]. Copyright IOP Publishing Ltd. CC BY 3.0.

A general expression for the Lorentz–Doppler profiles of components of HHSL obtained in paper [13] was:

$$I(w,\, b,\, \psi) = \int_0^{\infty} du_z f_z(\mathrm{u_z}) \int_0^{\infty} du_R f_R(u_R) \int_0^{\pi} (d\phi/\pi) g(\psi,\, \phi) \delta[w - \mathrm{u}_z \cos\psi - u_R(b + \sin\psi\cos\phi)]. \tag{9.26}$$

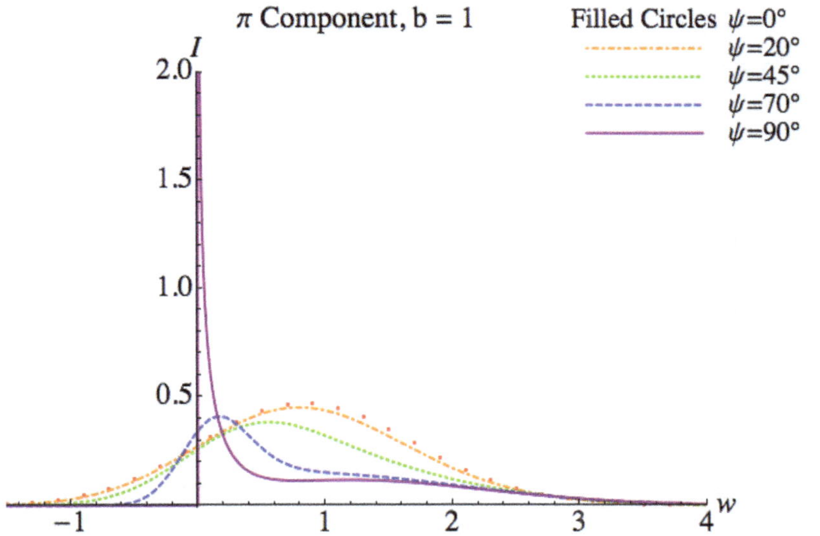

Figure 9.4. Same as in figure 9.2, but for $b = 1$. Reproduced from [20]. Copyright IOP Publishing Ltd. CC BY 3.0.

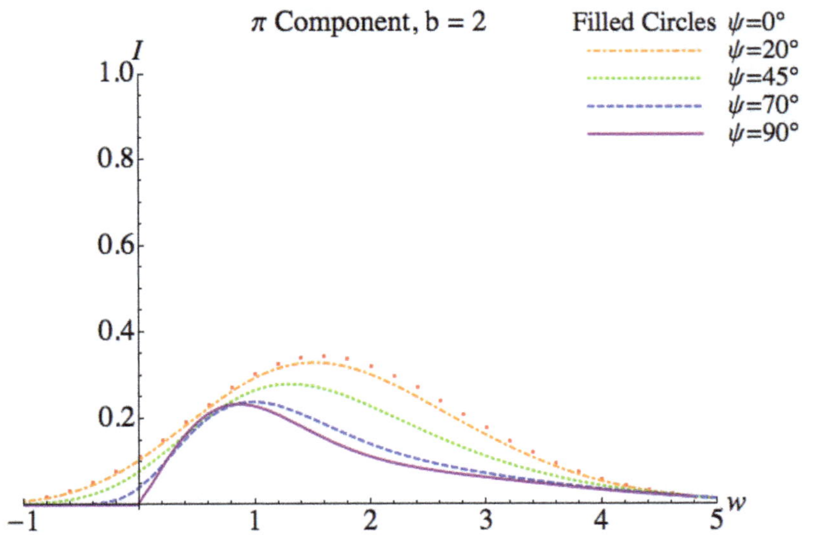

Figure 9.5. Same as in figure 9.2, but for $b = 2$. Reproduced from [20]. Copyright IOP Publishing Ltd. CC BY 3.0.

In equation (9.26),

$$f_z(u_z) = \frac{1}{\sqrt{\pi}} e^{-u_z^2}, f_R(u_R) = 2 \quad u_R e^{-u_R^2}, 0 < \psi < \frac{\pi}{2}, \tag{9.27}$$

where $g(\psi, \phi)$ are directional factors for π- and σ- components, as follows:

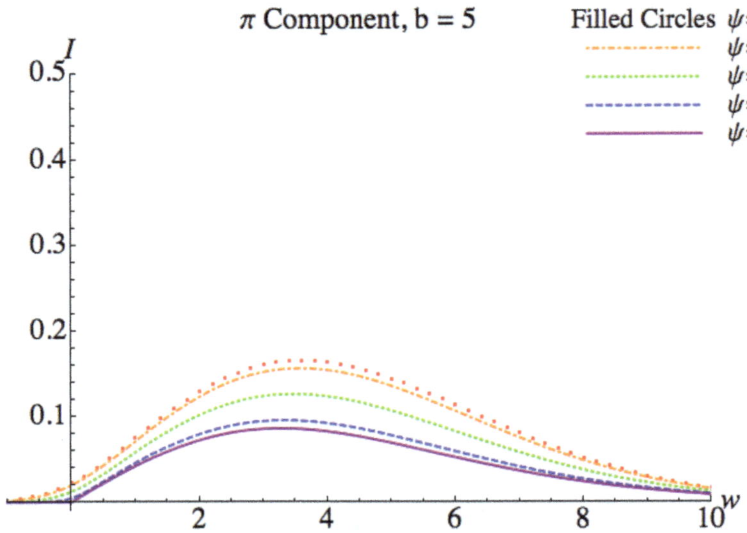

Figure 9.6. Same as in figure 9.2, but for $b = 5$. Reproduced from [20]. Copyright IOP Publishing Ltd. CC BY 3.0.

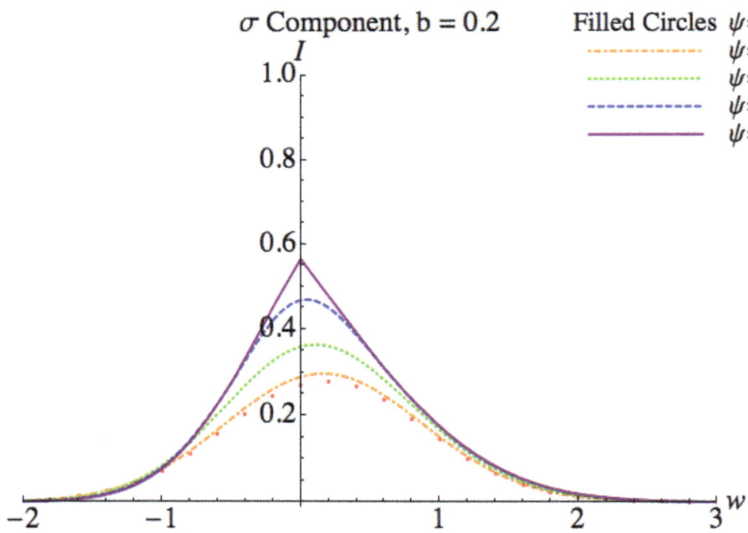

Figure 9.7. Calculated Lorentz–Doppler profiles of σ-components of highly-excited hydrogen or deuterium spectral lines for the scaled magnetic field $b = 0.2$ at five different angles of the observation. Reproduced from [20]. Copyright IOP Publishing Ltd. CC BY 3.0.

$$g_\pi(\psi) = 1 - \sin^2\psi\sin^2\phi,\ \ g_\sigma(\psi) = \frac{1}{2}(1 + \sin^2\psi\sin^2\phi). \tag{9.28}$$

Further in equation (9.26), $f_R(u_R)$ and $f_z(u_z)$ are the two-dimensional and the one-dimensional Maxwell distributions of the scaled atomic velocity $\mathbf{u} = \mathbf{v}/v_T$, respectively.

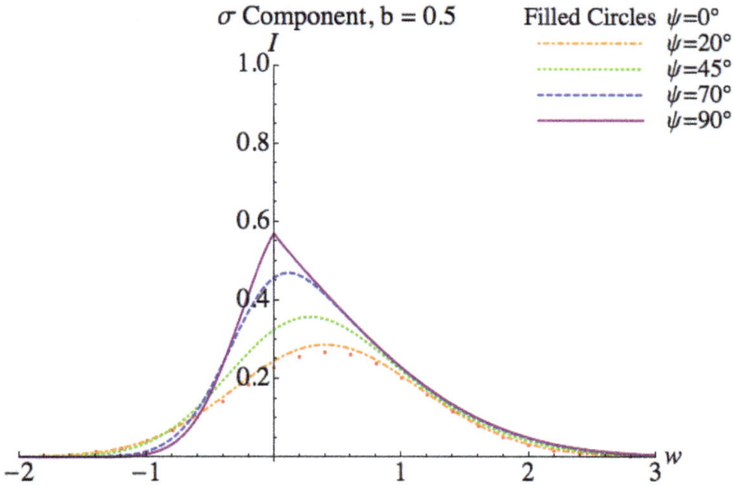

Figure 9.8. Same as in figure 9.7, but for $b = 0.5$. Reproduced from [20]. Copyright IOP Publishing Ltd. CC BY 3.0.

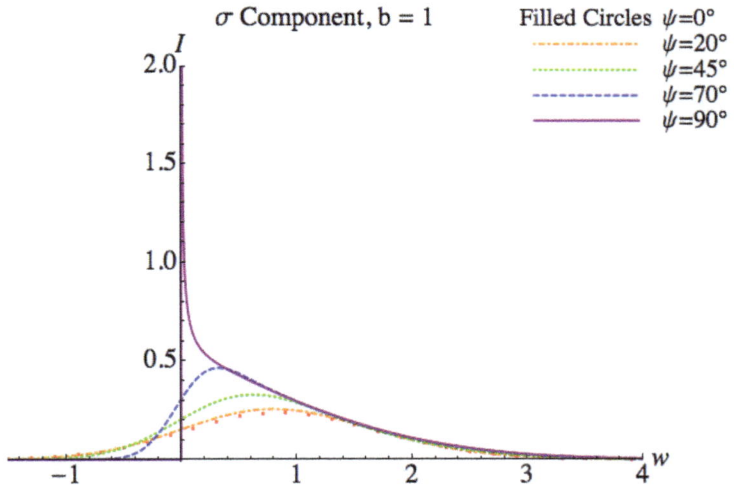

Figure 9.9. Same as in figure 9.7, but for $b = 1$. Reproduced from [20]. Copyright IOP Publishing Ltd. CC BY 3.0.

In paper [20] we performed analytically two (out of three) integrations in equation (9.26) to obtain:

$$I(w,\ b,\ \psi) = \frac{e^{-\frac{w^2}{\cos^2\psi}}}{\pi^{\frac{3}{2}}\cos\psi}\int_0^{\pi}\frac{g(\psi,\ \ \phi)}{[a(b,\ \psi,\ \phi)]^{\frac{3}{2}}} \left\{\sqrt{a(b,\ \psi,\ \phi)} + \sqrt{\pi}\ \ c(w,\ b,\ \psi,\ \phi)\ \ e^{\frac{c(w,\ b,\ \psi,\ \phi)^2}{a(b,\ \psi,\ \phi)}}\left[1 + \mathrm{Erf}\frac{c(w,\ b,\ \psi,\ \phi)}{\sqrt{a(b,\ \psi,\ \phi)}}\right]\right\}d\phi \tag{9.29}$$

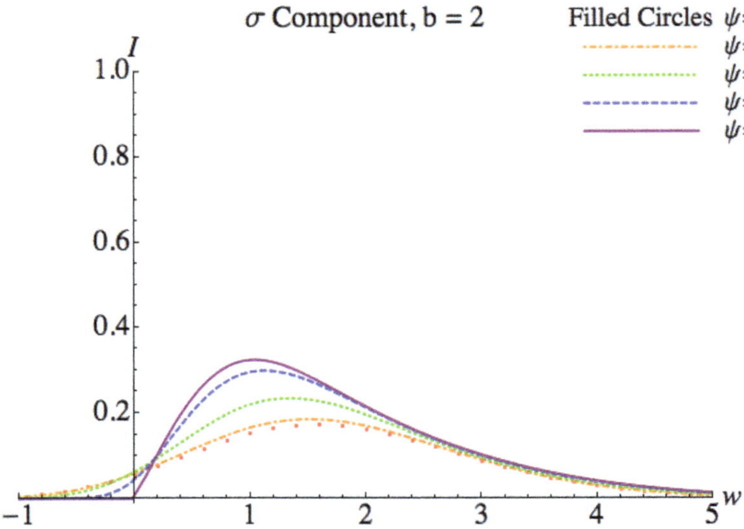

Figure 9.10. Same as in figure 9.7, but for $b = 2$. Reproduced from [20]. Copyright IOP Publishing Ltd. CC BY 3.0.

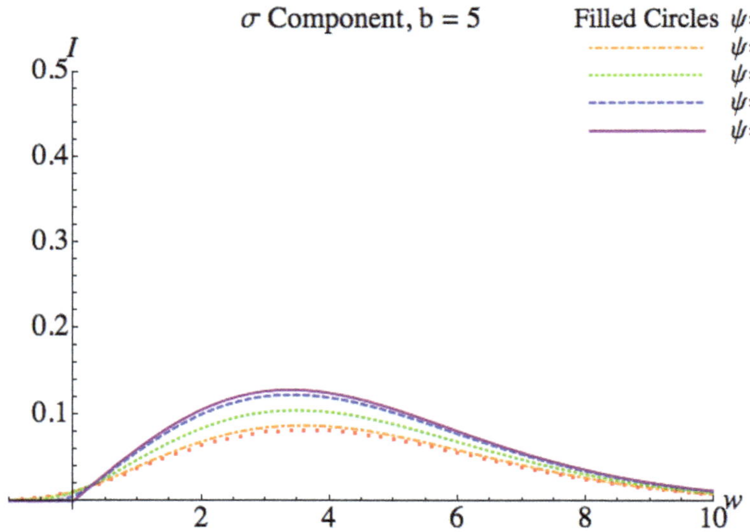

Figure 9.11. Same as in figure 9.7, but for $b = 5$. Reproduced from [20]. Copyright IOP Publishing Ltd. CC BY 3.0.

where

$$a(b, \psi, \phi) = 1 + \frac{(b + \sin\psi \cos\phi)^2}{\cos^2\psi}, \quad c(w, b, \psi, \phi) = w\,\frac{b + \sin\psi \cos\phi}{\cos^2\psi}. \tag{9.30}$$

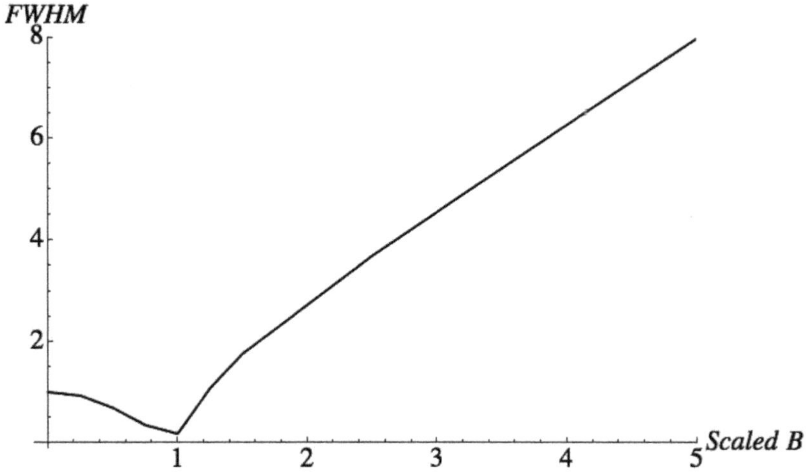

Figure 9.12. The calculated full-width at half-maximum (FWHM) of the hydrogen or deuterium Lyman-beta line observed perpendicular to the magnetic field **B**, while a linear polarizer is parallel to **B**. The scaled magnetic field is the ratio of the Lorentz-field shift to the Doppler shift. The FWHM is in units of the Doppler half-width at half-maximum. Reproduced from [20]. Copyright IOP Publishing Ltd. CC BY 3.0.

Figures 9.2–9.6 display the corresponding Lorentz–Doppler profiles of π-components of HHSLs. Each figure demonstrates profiles for five different angles of the observation (in degrees): 0, 20, 45, 70, and 90. Figures 9.2–9.6 are distinguished from each other by the chosen value of the scaled magnetic field: $b = 0.2, 0.5, 1, 2$, and 5.

Figures 9.7–9.11 display the analogous series of Lorentz–Doppler profiles for σ-components of HHSLs.

From figures 9.7 and 9.11, corresponding to the relatively strong magnetic field $b = 5$, one can see that, as the angle of the observation decreases from 90° to 70°, the suppression of the π-component, while decreasing, is still noticeable. However, as the angle of the observation reaches 45°, the suppression disappears. So, the suppression of the π-components (relative to the σ-components) takes place at the perpendicular or close to the perpendicular direction of observation, but rapidly vanishes at lower angles of the observation.

Another *counterintuitive* result is as follows. For observations perpendicular to the magnetic field, the width of the Lorentz–Doppler profiles exhibits a non-monotonic behavior with respect to the scaled magnetic field b. As b rises from zero, the width first diminishes, then reaches the minimum at $b = 1$ (that is, when the Doppler shift is equal to the shift in the Lorentz field), and then increases. This is illustrated in figure 9.12 by the example of the Lyman-beta line.

References

[1] Wroblewski D 1992 *Intern. Rev. Atom. Mol. Phys.* **No. 257** (New York: AIP) p 121

[2] Abramov V A and Lisitsa V S 1977 *Sov. J. Plasma Phys.* **3** 451

[3] Bychkov S S, Ivanov R S and Stotskii G I 1987 *Sov. J. Plasma Phys.* **13** 769

[4] Derevianko A and Oks E 1994 *Phys. Rev. Lett.* **73** 2059

[5] Ispolatov Y and Oks E 1994 *J. Quant. Spectrosc. Radiat. Transfer* **51** 129
[6] Oks E 2006 *Stark Broadening of Hydrogen and Hydrogenlike Spectral Lines in Plasmas: The Physical Insight* (Oxford: Alpha Science International)
[7] Griem H R 1974 *Spectral Line Broadening by Plasmas* (New York: Academic)
[8] Derevianko A and Oks E 1997 *Rev. Sci. Instrum.* **68** 998
[9] Oks E 2023 *Int. Rev. At. Mol. Phys.* **14** 43
[10] Sholin G V and Oks E 1973 *Sov. Phys. Doklady* **18** 254
[11] Oks E and Sholin G V 1976 *Sov. Phys. Tech. Phys.* **21** 144
[12] Oks E 2017 *Diagnostics of Laboratory and Astrophysical Plasmas Using Spectral Lineshapes of One-, Two-, and Three-Electron Systems* (Singapore: World Scientific)
[13] Oks E 2015 *J. Quant. Spectr. Radiat. Transfer* **156** 24
[14] Welch B L, Griem H R, Terry J, Kurz C, LaBombard B, Lipschultz B, Marmar E and McCracken J 1995 *Phys. Plasmas* **2** 4246
[15] Brooks N H, Lisgo S, Oks E, Volodko D, Groth M and Leonard A WDIII-D Team 2009 *Plasma Phys. Rep.* **35** 112
[16] Oks E 2012 *Atomic Processes in Basic and Applied Physics* ed V Shevelko and H Tawara (Heidelberg: Springer) ch 15
[17] Oks E, Bengtson R D and Touma J 2000 *Contrib. Plasma Phys.* **40** 158
[18] Feldman U and Doschek G A 1977 *Astrophys. J.* **212** 913
[19] Galushkin Y I 1970 *Sov. Astron* **14** 301
[20] Sanders P and Oks E 2017 *J. Phys. Commun.* **1** 055011

Chapter 10

Polarization and directional effects in the laser-induced fluorescence from plasmas and their applications

Active plasma diagnostic methods, that is, the methods using lasers, enlarge to capabilities spectroscopic diagnostics of Oscillatory Electric Fields (OEFs) in plasmas. These methods enable good spatial and temporal resolutions that most passive methods lack.

In the active spectroscopic methods for diagnosing the OEF $\overrightarrow{E_M}(t)$, atoms in plasmas are subjected simultaneously to $\overrightarrow{E_M}(t)$ and to the laser field $\overrightarrow{E_L}(t) = \overrightarrow{E_{0L}} \cos \omega_L t$. Typically, $\omega_M \ll \omega_L$, while the frequency ω_L is in resonance (or nearly in resonance) with the atomic transition frequency between the upper and lower energy levels involved in the radiative transition.

In paper [1], which we follow in this chapter, it was described how laser-aided spectroscopic methods can improve the diagnostics of OEFs in plasmas of magnetic fusion machines.

10.1 Absorption spectra of atoms not having permanent dipole moments

The interaction of a radiating atom with the OEF $\overrightarrow{E_M}(t)$ is controlled by the parameter

$$R = |d_{\alpha'\alpha''} E_{0M}/[2\hbar(\omega_{\alpha'\alpha''} \pm \omega_M)]|. \tag{10.1}$$

In equation (10.1), $\omega_{\alpha'\alpha''}$ is the separation between the closely spaced levels α', α'', while $d_{\alpha'\alpha''}$ is the corresponding dipole matrix element. For the relatively weak field

doi:10.1088/978-0-7503-6285-6ch10

$\overrightarrow{E_M}(t)$), one has $R \ll 1$, so that the field $\overrightarrow{E_M}(t) = \overrightarrow{E_{0M}} \cos \omega_M t$ can be determined experimentally by utilizing the fact that the excitation rate for the two-quantum process (one quantum of the laser field and one quantum of the OEF) is proportional to E_{0M}^2. One of the examples is the scheme presented in figure 10.1. It is assumed that $\omega_{23} \ll \omega_{21}$ and the single-photon transition $1 \leftrightarrow 2$ is considered to be forbidden, while the single-photon transitions $1 \leftrightarrow 3$ and $2 \leftrightarrow 3$ are considered to be allowed.

In frames of the perturbation theory, the corresponding excitation rates for the two-quantum transitions are:

$$\begin{aligned} W_{1\to 2}^{(+)} &= \{E_{0L}^2 E_{0M}^2 d_{13}^2 d_{32}^2 / [8\hbar^2(\omega_{32} + \omega_M)^2]\} L(\omega_{21} - \omega_L - \omega_M, \gamma), \\ W_{1\to 2}^{(-)} &= \{E_{0L}^2 E_{0M}^2 d_{13}^2 d_{32}^2 / [8\hbar^2(\omega_{32} - \omega_M)^2]\} L(\omega_{21} - \omega_L + \omega_M, \gamma). \end{aligned} \tag{10.2}$$

Here γ is the width (the full-width at half-maximum) of the line, while $L(\omega, \gamma) \equiv (\gamma/2)/(\omega^2 + \gamma^2/4)$, The quantities $W_{1\to 2}^{(+)}$and $W_{1\to 2}^{(-)}$ correspond to the processes depicted in figures 10.1(a) and (b), respectively. For the single-photon transition $1 \to 3$ (induced by the laser field) the excitation rate is:

$$W_{1\to 3} = [d_{13}^2 E_{0L}^2/(2\hbar^2)] L(\omega_{21} - \omega_L, \gamma). \tag{10.3}$$

The ratios $W_{1\to 2}^{(-)}/W_{1\to 3}$ and $W_{1\to 2}^{(+)}/W_{1\to 3}$ are proportional to E_{0M}^2 and can be utilized for measuring E_{0M}. For this purpose, it is sufficient to scan the laser frequency near the frequencies ω_{21} and ω_{31} and observe the fluorescence signals at transitions from the levels 2 and 3.

In the opposite situation where $R \sim 1$ or R 1, the best analytical formalism seems to be the formalism of quasienergy states (QES). It relates to the basis of the following wave functions (see, e.g. [2]):

$$\Phi_n(\vec{r}, t) = \exp(-i\varepsilon_n t/\hbar) \sum_k \exp(-ik\omega_M t)\phi_{nk}(\vec{r}). \tag{10.4}$$

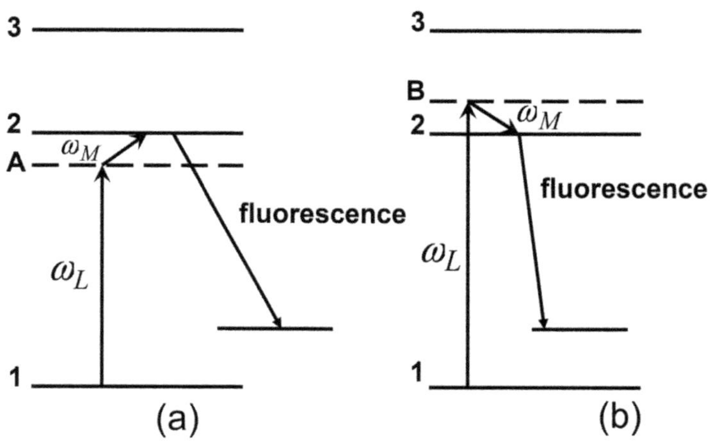

Figure 10.1. Example of the excitation scheme. (a) The excitation of the transition $1 \to 2$ is by the absorption of an OEF quantum $\hbar\omega_M$ and a laser photon $\hbar\omega_L$; (b) the excitation of the transition $1 \to 2$ by the absorption of a laser photon $\hbar\omega_L$ and the emission of an OEF quantum $\hbar\omega_M$. Reproduced from [1] with permission.

Here ε_n is the quasienergy of the level n, the term 'quasienergy states' being introduced in papers [3, 4].

The excitation of the transition from some level b to some level a requires the following resonance condition to be met

$$\omega_L \approx (\varepsilon_a - \varepsilon_b)/\hbar + k\omega_M, \quad k = 0, \pm 1, \pm 2, \ldots \tag{10.5}$$

In equation (10.5), ε_a and ε_b are the quasienergies of the levels a and b, respectively. The excitation rate for the transition $b \rightarrow a$ can be calculated by utilizing equation (10.3) on substituting the dipole matrix element d_{13} by the effective dipole matrix element $D_{ba}^{(k)} = \langle \varphi_b | d | \phi_{ak} \rangle$ (assuming that the OEF perturbs only the upper state a):

$$W_{b\rightarrow a} = \left[\left| \left\langle \varphi_b | d | \phi_{ak} \right\rangle \right|^2 E_{0L}^2/(2\hbar^2) \right] L((\varepsilon_a - \varepsilon_b)/\hbar - \omega_L + k\omega_M, \gamma). \tag{10.6}$$

10.2 Absorption spectra of atoms possessing permanent dipole moments

Now we consider atoms having permanent dipole moments ($d_{aa} \equiv \langle \varphi_a | d | \varphi_a \rangle \neq 0$), such as, for example, hydrogen atoms. The wave function of a QES of the atom under the OEF $\overrightarrow{E_M}(t) = \overrightarrow{E_{0M}} \cos \omega_M t$ can be represented as follows [5]:

$$\Psi_a(\overrightarrow{r}, t) = \exp[-\mathrm{i}d_{\mathrm{aa}} E_{0M} \sin \omega_M t/(\hbar\omega_M)] \varphi_a(\overrightarrow{r}). \tag{10.7}$$

If the frequency ω_L of the laser radiation is resonant (or nearly resonant) to the frequency of the transition $b \leftrightarrow a$, then using the Fourier series expansion in equation (10.7) and employing the relation $\exp(-ix \sin \omega_M t) = \sum_{n=-\infty}^{+\infty} J_n(x) \exp(-in\omega_M t)$, where $J_n(x)$ are the Bessel functions, one obtains the following effective matrix element of the dipole moment between the levels a and b:

$$D_{ab}^{(k)} = J_k(\Delta\beta_{ab}) d_{ab}. \tag{10.8}$$

In equation (10.8), $\Delta\beta_{ab} \equiv (d_{aa} - d_{bb}) E_{0M}/(\hbar\omega_M)$. Then the increase of the population ΔN_a of the upper level a can be expressed as follows:

$$\begin{aligned} &\Delta N_a \equiv N_a - N_{a0} = 2^{-1} G_k (N_{b0} - N_{a0})/[1 + (\omega_{ab} + k\omega_M - \omega_L - \sigma_k)^2 \tau_{12}^2 + G_k], \\ &G_k \equiv 4 W_{ab}^2 J_k^2(\Delta\beta_{ab}) \tau_{12} \Gamma^{-1}, \quad \sigma_k = 2\omega_M^{-1} W_{ab}^2 g_k(\Delta\beta_{ba}), \\ &g_k(\nu) = \sum_{r=1}^{\infty} [J_{k-r}^2(\nu) - J_{k+r}^2(\nu)]/r, \quad W_{ab} \equiv d_{ab} E_{0L}/(2\hbar), \quad k = 0, \pm 1, \pm 2, \ldots \end{aligned} \tag{10.9}$$

In equation (10.9), $\Gamma^{-1}$and τ_{12} are the longitudinal and transverse relaxation times, respectively. The transverse relaxation time τ_{12} is inversely proportional to the so-called homogeneous width (the impact width) of the spectral line $a \rightarrow b$. As for the parameter Γ, it controls the relaxation rate of the levels a and b.

If the laser field is relatively weak, equation (10.9) can be simplified as follows:

$$\Delta N_a = \frac{d_{ab}^2 E_{0L}^2}{2\hbar^2} \frac{J_k^2(\Delta\beta_{ab})\tau_{12}\Gamma^{-1}(N_{b0} - N_{a0})}{1 + (\omega_{ab} + k\omega_M - \omega_L)^2\tau_{12}^2}, \quad k = 0, \pm 1, \pm 2, \ldots \qquad (10.10)$$

Equation (10.10) corresponds to Blokhinzew's spectrum [5]: the intensity of the satellites at the frequencies $\omega_L = \omega_{ab} + k\omega_M$ are proportional to $J_k^2(\Delta\beta_{ab})$.

From equation (10.9) it can be seen that generally the spectrum of the atomic absorption of the laser radiation should exhibit a set of satellites at the frequencies

$$\omega_L = \omega_{ab} + k\omega_M - \sigma_k, \quad k = 0, \ \pm 1, \ \pm 2, \ldots \qquad (10.11)$$

The half-width of the kth satellite can be written as

$$\Delta\omega_{k,1/2} = 2(1 + G_k)^{1/2}/\tau_{12}. \qquad (10.12)$$

The term σ_k in the denominator of the expression for ΔN_a in equation (10.9) is caused by the dynamic Stark shift of the quasienergy levels. It depends on both the OEF $\overrightarrow{E_M}(t)$ and the laser field. The above formula for σ_k was derived in papers [6, 7].

From equation (10.9) it can be seen how to measure the amplitude of the microwave field E_{0M} in plasmas. The authors of paper [1] described the following details.

'For this purpose one should record the wavelength-integrated intensity of LIF $I_f \propto \Delta N_2$ during the transition from the upper level a to one of the lower levels versus the laser field intensity $I_L \propto E_{0L}^2$: $I_f^{-1} \propto (1 + I_{k,\,\mathrm{satur}} I_L^{-1})$, $I_{k,\,\mathrm{satur}} I_L^{-1} \equiv G_k^{-1}(E_{0M})$. Here k is the number of microwave quanta involved in the resonance. The amplitude E_{0M} can be measured in two ways. In the first way, one could measure the ratio of slopes of experimental dependences $I_f^{-1}(I_L^{-1})$ at two different values of $k = 0$ ($k = k'$, and $k = k''$), and use the dependence of the ratio $G_{k'}/G_{k''}$ on E_{0M}. In the second way, one could use the dependence of the dynamic Stark shift σ_k on the parameters of the OEF $\overrightarrow{E_M}(t)$. By scanning the laser frequency around the frequency ω_{ab} and by recording peaks of the LIF intensity, one could tune to multi-quantum resonances (8) corresponding to two different indices $k = k_1$, and $k = k_2$. The laser frequency should be kept constant. In this situation one would have $\omega_{L,1} = \omega_{ab} + k_1\omega_M - \sigma_{k_1}$, $\omega_{L,2} = \omega_{ab} + k_2\omega_M - \sigma_{k_2}$.

Since ω_M is known, one would determine the dynamic Stark shifts σ_{k_1} and σ_{k_2}. Then, using the fact that the ratio $\sigma_{k_2}/\sigma_{k_1}$ depends on E_{0M} but does not depend on E_{0L}, one could finally find the OEF amplitude E_{0M}.

Finally, the authors of paper [1] analyzed the case where a beam of hydrogen or deuterium atoms travels through the plasma of a magnetic fusion machine, the plasma containing OEF $\overrightarrow{E_M}(t)$. The hydrogen atoms are under the electric field $\vec{\varepsilon}(t) = \overrightarrow{E_b} + \overrightarrow{E_M}(t)$, where $\overrightarrow{E_b} = \overrightarrow{V_b} \times \vec{B}/c$ is the Lorentz electric field. The latter field $\overrightarrow{E_b}(t)$ produces the static Stark splitting of the hydrogen energy levels. The beam velocity can be such chosen as for the Stark splitting to be in a resonance with the frequency $\omega_M$of the OEF $\overrightarrow{E_M}(t)$. The resonance condition can be written as

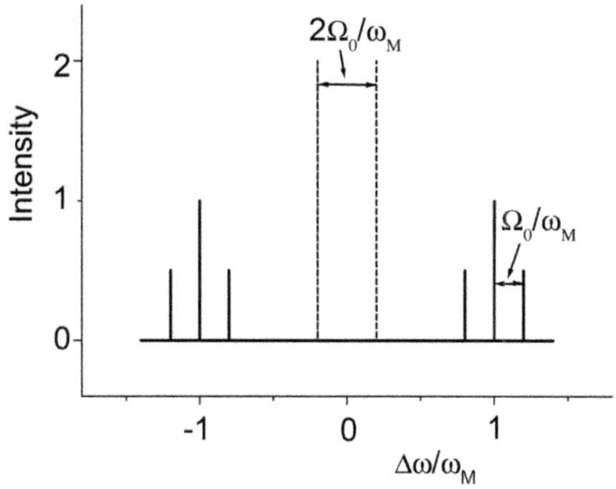

Figure 10.2. Spectrum of the hydrogen line L_α in two mutually perpendicular electric fields: the oscillatory field $\overrightarrow{E_M}(t) = \overrightarrow{E_{0M}} \cos \omega_M t$ and the Lorentz field $\overrightarrow{E_b}$—under the conditions of the resonance (10.10) for the energy level of the principal qurntum number $n = 2$. The direction of the observation is perpendicular to both electric fields. The spectrum is shown for two orientations of the linear polarizer: parallel to $\overrightarrow{E_b}$ (solid line) and parallel to $\overrightarrow{E_M}(t)$ (dashed line). The corresponding Rabi frequency, to which the splitting is proportional, is $\Omega_0 = \frac{3\hbar}{2me} E_{0M}$. Reproduced from [1] with permission.

$$\omega_M = \frac{3\hbar}{2me} n E_b, \tag{10.13}$$

where n is the principal quantum number.

If the resonance condition (10.13) is satisfied, then the probability to find the hydrogen atom is in a Stark sublevel of the principal quantum number n oscillates at the Rabi frequency, the latter being proportional to the amplitude of the OEF. The consequence is an additional splitting of the hydrogen spectral lines, corresponding to the radiative transition involving the level of the principal quantum number n as the upper or the lower level. The result is the additional splitting of the hydrogen spectral line, the additional splitting being proportional to the amplitude of the OEF.

Figure 10.2 displays the spectrum of the hydrogen line L_α in the condition of the resonance (10.13) for the case where the upper level is $n = 2$. Recording the additional splitting can serve as the sensitive method to measure the amplitude of the OEF in magnetic fusion plasmas. For arranging the resonance (10.13), the experimentalists should vary the energy of the beam.

Figure 10.2 demonstrates the clear difference between the spectra of the hydrogen line [under the resonance (10.10)] in the two mutually perpendicular polarizations.

References

[1] Gavrilenko V P and Oks E 2011 *Intern. Rev. Atom. Mol. Phys.* **2** 35
[2] Delone N B and Krainov V P 1985 *Atoms in Strong Light Fields* (Berlin: Springer)
[3] Zel'dovich J B 1967 *Sov. Phys. JETP* **24** 1006
[4] Ritus V I 1967 *Sov. Phys. JETP* **24** 1041

[5] Blochinzew D I 1933 *Phys. Z. Sow. Union* **4** 501
[6] Gavrilenko V P and Oks E 1995 *Phys. Rev. Lett.* **74** 3796
[7] Gavrilenko V P and Oks E 1995 *J. Phys. B: At. Mol. Opt. Phys.* **28** 1433

www.ingramcontent.com/pod-product-compliance
Ingram Content Group UK Ltd.
Pitfield, Milton Keynes, MK11 3LW, UK
UKHW051850100926
8556IPUK00005B/558

9 780750 362832